Clinical Engineering

Clinical Engineering
From Devices to Systems

Roberto Miniati, PhD., M.Sc.

Department of Information Engineering – Biomedical Lab
Via di Santa Marta, Firenze, Italy

Ernesto Iadanza, Ph.D., M.Sc., CE

Chairman, Clinical Engineering Division of the International Federation
for Medical and Biological Engineering (IFMBE),
Chairman, Education and Accreditation Committee of the International Union
for Physical and Engineering Sciences in Medicine (IUPESM)
Department of Information Engineering – Università di Firenze, Italy

Fabrizio Dori, M.Sc.

Department of Information Engineering – Biomedical Lab
Via di Santa Marta, Firenze, Italy; Regional Health Technology Department,
Tuscany, Italy

ELSEVIER

AMSTERDAM • BOSTON • HEIDELBERG • LONDON
NEW YORK • OXFORD • PARIS • SAN DIEGO
SAN FRANCISCO • SINGAPORE • SYDNEY • TOKYO
Academic Press is an imprint of Elsevier

Academic Press is an imprint of Elsevier
125, London Wall, EC2Y 5AS.
525 B Street, Suite 1800, San Diego, CA 92101-4495, USA
225 Wyman Street, Waltham, MA 02451, USA
The Boulevard, Langford Lane, Kidlington, Oxford OX5 1GB, UK

ISBN: 978-0-12-803767-6

British Library Cataloguing-in-Publication Data
A catalogue record for this book is available from the British Library.

Library of Congress Cataloging-in-Publication Data
A catalog record for this book is available from the Library of Congress.

For Information on all Academic Press publications
visit our website at www.elsevier.com

Typeset by MPS Limited, Chennai, India
www.adi-mps.com

Printed and bound in the United States

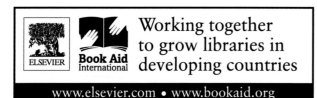

Working together
to grow libraries in
developing countries

www.elsevier.com • www.bookaid.org

Publisher: Joe Hayton
Acquisitions Editor: Fiona Geraghty
Editorial Project Manager: Natasha Welford
Production Project Manager: Lisa Jones
Designer: Victoria Pearson

Contents

List of Contributors

Zhivko Bliznakov   Biomedical Technology Unit, Department of Medical Physics, Faculty of Medicine, University of Patras, Rio - Patras, Greece

Cari Borrás   Radiological Physics and Health Services, Washington DC, USA

Marcello Bracale   Department of Surgical Specialities, Nephrology Università degli studi di Napoli Federico II, Napoli, Italy

Umberto Bracale   Department of Surgical Specialities, Nephrology Università degli studi di Napoli Federico II, Napoli, Italy

Saide Jorge Calil   Faculty of Electrical Engineering and Computing, University of Campinas, Campinas, São Paulo, Brazil; Clinical Engineering Division, IFMBE

Rossana Castaldo   School of Engineering, University of Warwick, UK

Michael Craven   University of Nottingham, Nottingham, UK

Zhou Dan   Medical Management Department, PLA General Hospital and Provost, PLA Medical College; Clinical Engineering Branch, Medical Doctor Association; Clinical Engineering Branch, Chinese Society of Biomedical Engineering; Medical Engineering Branch, Chinese Medical Association, Beijing, China

Yadin David   Biomedical Engineering Consultants, LLC; University of Texas School of Public Health, Houston, TX, USA

Fabrizio Dori   Department of Information Engineering, Biomedical Lab, Via di Santa Marta, Firenze, Italy; Regional Health Technology Department, Tuscany, Italy

Francesco Frosini   Department of Information Engineering, Biomedical Lab, Via di Santa Marta, Firenze, Italy

Gabriele Guidi   Department of Information Engineering, Biomedical Lab, Firenze, Italy

Torsten Gruchmann   Use-Lab GmbH, Steinfurt, Germany

Fred Hosea   Kaiser Permanente, Clinical Technology (ret.), Convivia, President

Ernesto Iadanza   Chairman, Clinical Engineering Division of the International Federation for Medical and Biological Engineering (IFMBE), Chairman, Education and Accreditation Committee of the International Union for Physical and Engineering Sciences in Medicine (IUPESM), Department of Information Engineering - Università di Firenze, Italy

Alessio Luschi   Department of Information Engineering, Biomedical Lab, Firenze, Italy

Roberto Miniati   Department of Information Engineering, Biomedical Lab, Via di Santa Marta, Firenze, Italy

Mario Medvedec   University Hospital Centre Zagreb, Department of Nuclear Medicine and Radiation Protection, Zagreb, Croatia

Paolo Melillo   Dipartimento di Oftalmologia, Seconda Università degli studi di Napoli, Naples, Italy

Frank R. Painter   Clinical Engineering, University of Connecticut, Storrs, CT, USA

Nicolas Pallikarakis   Biomedical Technology Unit, Department of Medical Physics, Faculty of Medicine, University of Patras, Rio - Patras, Greece

Leandro Pecchia   School of Engineering, University of Warwick, UK; Healthcare Technology Assessment Division, International Federation for Medical and Biological Engineering (IFMBE), Paris, France

Mário Forjaz Secca   Department of Physics, Faculty of Science and Technology, Universidade Nova De LISBOA, Lisboa, Portugal

Elliot B. Sloane   Research Faculty, Villanova University Center for Enterprise Excellence in Technology, Villanova, PA, USA, and Executive Director, Center for Healthcare Information Research and Policy, Osprey, FL, USA

Heikki Teriö   Clinical Engineering, Karolinska University Hospital, Stockholm, Sweden

James O. Wear   Scientific Enterprises, North Little Rock, AR, USA

Acknowledgments

We would like to thank all the colleagues involved in this project, the book contributors for their great expertise and professionalism, the reviewers for increasing the book quality, and Elsevier for making this project a reality.

Mostly we want to thank you, Reader — as we believe that sharing knowledge is the only key to sustainable progress.

Roberto, Ernesto, and Fabrizio

Introduction

With the increase of scientific knowledge and technological progress in healthcare, a new vision and a new approach are now requested to effectively design and manage numerous technologies, nowadays an inseparable part of clinical activity in various healthcare settings including home. This indissoluble complex system composed by users, technologies, patients, and structures can be defined as a "Clinical System." It is a new working domain for technology professionals within current and future healthcare.

As clearly suggested by the books title *Clinical Engineering: From Devices to Systems* the aim of this project is to provide the most innovative experiences regarding the new models, methods, and challenges faced by Professionals, Technology Managers, Manufacturers, Advisors, and Academics in the area of Clinical Engineering, Medical Physics, and Health Economics.

Hence, the cases reported in the book have the ambition to support the reader by providing real experiences from the most important hospitals and universities worldwide and suggesting solutions on how this complex matter can be managed, especially when personal experience makes a difference and many disciplines, such as physics and chemistry, engineering (mechanics and electronics), information technology, economics, and medical notions, are necessary to succeed.

The book, with the contributions of the best specialists from all over the world, starts with the history of Clinical Engineering and its evolution in order to set the proper background and the current scientific borders. Then it brings the focus to a worldwide level, to find out the meaning of being a Clinical Engineer in various markets and scenarios. Direct experiences coming from developing countries and evolving and consolidated markets are included, such as examples from Mozambique, China, and Europe.

Afterwards, a list of ordinary tasks involving technology managers and designers is presented: first, real cases related to everyday management activity of hospital technology, including both common devices and complex systems; second, specific cases where technology provides support to other professionals, reporting support systems for clinical decisions or specific methods applications for rapid evaluation of technology cost-effectiveness.

Next, the book focuses on risk, quality, and safety engineering applied to clinical systems. An integrated approach of quality and risk management is presented in order to highlight how important this synergy can be for reducing risk and increasing hospital efficiency. Then, methodological guidelines are provided on how to safely use, manage, and design technological systems or new technologies incorporating software. The last contribution of this section is a chapter highlighting the importance of hospital technology management for reducing disaster risk. Finally, the importance of human factor engineering for reducing clinical and technological risks within healthcare is described.

As a closing topic, the book provides an overview and specific experiences of careers of clinical engineers and their education before describing the existing and future professional certification programs for Clinical Engineers.

1

The Evolution of Clinical Engineering: History and the Role of Technology in Health Care

Saide Jorge Calil

FACULTY OF ELECTRICAL ENGINEERING AND COMPUTING, UNIVERSITY OF CAMPINAS, CAMPINAS, SÃO PAULO, BRAZIL
CLINICAL ENGINEERING DIVISION, IFMBE

There is today no precise data about how many health units are there in the world. As of 2014, the countries with the largest number of hospitals are China, India, Vietnam, Nigeria, Russia, Japan, Egypt, South Korea, Brazil, and United States. As of 2014, China alone had around 69,000 hospitals, according to Maps of the World (mapsofworld.com, 2015). Also, it is almost impossible to estimate how many clinical engineers are employed by health systems worldwide, since the requirements for this professional varies from country to country.

If one looks at professions such as civil, mechanical, and electrical engineering, for instance, the competences, activities, and the basic requirements are worldwide known. However, regarding clinical engineering, one of the major challenges to be dealt with is the definition of the basic requirements a person must attend to be considered a clinical engineer.

This chapter presents what is happening today with clinical engineering worldwide, how it evolved in some countries, the different activities developed by clinical engineers according to the world region he/she is working, and the resulting professional profile in relation to each country's health system requirements.

Despite the lack of bibliography, probably medical equipment maintenance was born when the human warfare became more organized. By the time, doctors need help from "technical people" to sharpen instruments designed to withdraw arrow tips and amputate members from wounded warriors. Perhaps, they also needed some help to discuss and choose the best metal composition and shape to be used, without bending or loosing sharpness during "surgery."

What is called today clinical engineering originated just after 1950, when more complex technology started to be developed and used in the health system (Dyro, 2004).

Clinical engineering evolved according to the improvement and additional needs of the health system environment. Around 1979, Scott and Caceres (Scott, 1976; Caceres, 1981)

Clinical Engineering. DOI: http://dx.doi.org/10.1016/B978-0-12-803767-6.00001-5

separately identified a number of clinical engineering responsibilities. The combined list of responsibilities includes development and management of medical systems, education, maintenance, safety, clinical research and development, and analysis and development for the more effective patient care systems (Dolan, 2004). Betts (1983), in considering the changing role of clinical engineering in the 1980s, emphasized the need for management skills in addition to the requisite technical knowledge.

In Europe 1992, clinical engineering was concentrated on technology management and advisory regarding acquisition, training, and technological evaluation for equipment incorporation (Bravar, 2010). Later, in 2003, to the former defined activities risk management was added. Also, to the task of technology management, the responsibility over financial management of technologies was included.

Table 1.1 summarizes, in general terms, the activities for clinical engineering in the seventies and the additional activities that were included from the nineties till the present days.

In 2005, the Clinical Engineering Division of the International Federation of Medical and Biological Engineering (CED/IFMBE) carried out a worldwide survey to identify clinical engineers and characterize their performed activities. This survey included questions regarding name, age, email address, academic background, and years of experience within the clinical engineering area. For the activities, the employer, job position, and job activities were asked. To avoid misunderstandings by the respondents, regarding the activities they were currently developing, a glossary about the meaning of each activity was included, using the same definitions adopted by the American College of Clinical Engineering (ACCE, 2001).

From the 559 questionnaires that were responded, 54% were from Latin America, 27% from Europe, 10% from the United States/Canada, 8% from Asia, and 1% from Africa (Figure 1.1).

Table 1.1   Set of Activities Required to Clinical Engineering in the Seventies and in the Present Days

1970–1980	1990–2012
Equipment Management	Technology Management
Safety	Risk Management
Acquisition	Equipment incorporation
Education	Education
	Disaster preparedness
	Cost control (Total Cost Ownership)
	Technology Management
	Tele-medicine (Home care)
	Project Management
	Contract Management
	Mobile Health (events, transport)
	Quality Management
	System Management
	Information Technology

The results for the question regarding the present developed activity (Figure 1.2) show that Technology Management (60.8%) and Service Delivery (60.6%) are the mainly practiced activities. Both results are quite similar analyzing the responses from the four regions. Importantly, to answer this question, the clinical engineer could choose as many answers that identified his/her kind of activity in the working place.

Other activities, however, vary quite significantly according to each region (Figure 1.3). While in the Latin American, European, and US/Canadian regions the clinical engineers are very much involved with "Education" (around 53% in general), only 36.4% from the Asian region are involved with this activity, where the "Academia" is the second main employer (16%).

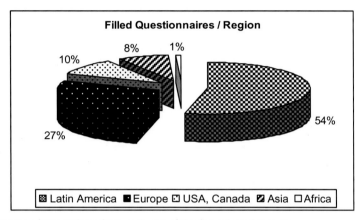

FIGURE 1.1 Percentage of respondents from each one of the five selected regions.

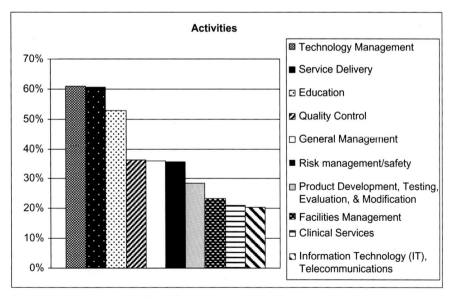

FIGURE 1.2 Activities that are practiced by clinical engineers worldwide.

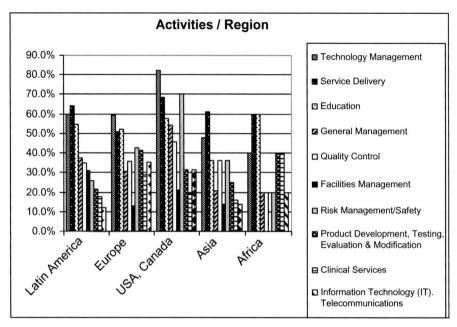

FIGURE 1.3 Activities that are practiced by clinical engineers within the five selected regions.

The reason for such event is difficult to explain since the employer's full address, for further explanation if necessary, was not asked in the questionnaire.

Around 70% of the clinical engineers from the US/Canadian region declared to be involved with "Risk Management/Safety" activities, while about 42%, 36%, and 26% of the respondents in the European, Asian, and Latin American regions, respectively, are involved with such activity.

"Information Technology" is other activity that varies quite significantly according to the regions analyzed. While in the European (35.3%) and US/Canadian (31.6%) regions several respondents declared to be involved with it, in the Latin American (12%) and Asian (13.6%) regions, just a few clinical engineers have such activity.

The analysis of this survey showed that complex clinical engineering activities, such as information technology and risk management, are concentrated on more developed regions. It does not mean that there are no needs for such activity in less developing regions as many hospitals acquire sophisticated medical technologies and indeed require the use of additional knowledge. However, the level of training courses (quality and contents) for clinical engineers on such less developing regions does not attend to such needs.

Another conclusion was that clinical engineering activities are still evolving in different ways and directions, not only according to the world regions but also to each country. As an example, while some European countries, as well as United States/Canada, are discussing the best way to implement information technology and the use of risk management tools within the clinical engineering activities, other regions are still discussing rules for equipment acceptance and installation and asking for training courses on the basic tools for medical equipment management.

The same can be said regarding the clinical engineering activities and the biomedical technician activities. While in some countries such differentiation is quite well defined, in several other countries clinical engineers work as the "repair person" for medical equipment. Hence, according to the country, even neighboring countries, the understanding of what is a clinical engineer is quite diverse. While in one country clinical engineers are known as the problem solver, the risk manager, the investigator, and the standard advisors, in other countries they are just recognized as "the boys for maintenance."

One could think that clinical engineering evolution progressed according to the number of years it was created in the country. However, this is strongly subject to debate, and it is not confirmed. In many countries where you have maintenance groups for decades, clinical engineers are still the medical equipment maintenance managers and have little role in the processes for acquiring medical equipment or even involved on risk management. The majority of clinical engineers in several regions around the world have no knowledge about what tools can be used in the investigation of an adverse event.

The reason for such different evolution is still to be found but it may be possible to take some conclusions. Where the health system management is worried about patient safety programs, has a thorough cost control and quality control for medical equipment incorporations and bidding process, it is possible to notice a quite sophisticated set of activities for clinical engineers. In countries where most of the preoccupation is to have a cheaper maintenance, the clinical engineering activities go no further than medical equipment maintenance management.

It seems that the evolution of the clinical engineering as well as the sophistication of the tasks to be developed is highly dependent on the understanding of the health system about the priorities in the care its population. The more the health system cares about safety, equipment integration and interoperability, cost control, and technology evaluation, the more clinical engineers have to be prepared to face such requirement.

There is, however, another reason why in some countries clinical engineering is still developing basic activities. The paucity of scientific articles produced by clinical engineers is hardly found in other engineering professions. As a consequence, there are quite a few journals exclusively dedicated to clinical engineering. The consequence is that clinical engineers do not share their experience among themselves and do not feel compelled to learn or apply more complex activities. Most of them keep isolated in his/her work. Importantly, some biomedical engineering journals do not accept articles describing practical methodologies and results described by clinical engineers, regardless the merit.

Additionally, it is common knowledge that clinical engineers produce a reduced amount of articles. Such attitude does not improve the understanding about the role of clinical engineering in the health sector and, in general and consequently, hospital administrators have little information about the latent qualities and abilities of this profession.

Explaining What Is a Clinical Engineer

To explain clinical engineering for a person who is not familiar with the profession may be not so difficult. It can be said that it is the engineer who works in the healthcare service or in a hospital. However, explaining to this same person what are the activities of the clinical engineer and his/her role in the healthcare system may be a bit more complex matter and quite complicated to be described in a single phrase.

As already stated, the creation and evolution of clinical engineering was totally linked to the needs of the health area on each country. Also, depending on such needs at the time of its establishment as a profession, clinical engineering will carry a stigma for many years to come. To better explain such statement, let's take the example of the United States where the original needs for its creation was basically patient safety (Bronzino, 2003). The clinical engineering model developed over there has a lot to do with risk and financial management, as well as contract management and internal operations. They even produced several guidelines on such subjects that are helping clinical engineers all over the world. Eventually, they can deal with the supervision of in-house maintenance staff (Grimes, 2012). Over there, a biomedical equipment technician (BMET) is a professional who develops the medical/hospital equipment maintenance.

In other world regions, such as Latin America, clinical engineering was developed basically to manage the maintenance of medical equipment. Hence, clinical engineers are seen as the person who either do the medical/hospital equipment maintenance and/or manage it. In fact, clinical engineering in these regions was originated basically due to the need of lowering the prices charged by external companies (service or vendors), to make repair and, sometimes, preventive maintenance. This is today the stigma carried by the majority of clinical engineers, mainly the ones who are frightened to stand up and present their skills to perform other tasks to the hospital administration.

Going to Japan, clinical engineers not only develop the maintenance of complex equipment but also work as an active health staff member during specific medical procedures. Such activity involves a direct participation, with doctors and nurses, on surgeries such as cardiac valve implants, cardiac catheterization, blood purifications, and several others. Importantly, his/her participation is not as a standby person waiting to solve some operational difficulty but actually is a member of the surgical team dealing and handling with the equipment, accessories, and even the part to be implanted in the patient.

Another example of the development of different clinical engineering models is in the European continent, where some countries use clinical engineers to monitor and manage external maintenance service companies, while in other countries most hospitals have their own clinical engineering group but developing different tasks. A German clinical engineer is quite dedicated to risk programs; in United Kingdom, there are clinical engineering groups dealing with rehabilitation developments as well as medical equipment maintenance. Some UK groups included services for the Ministry of Health regarding testing and conformity with standards.

Not only the mission of clinical engineers is different depending on the region but also the name. Biomedical engineers and hospital engineers are also used for the engineer that works within the healthcare system.

Requirements to Be a Clinical Engineer

Requirements to work as clinical engineers also vary according to the country. Some countries require certification of competence to work in this area, while others have a voluntary certification and most of them have no requirement at all.

The certification system is also different from country to country; some require the clinical engineer to be interviewed and approved in a set of written exams. Some require the filling of a form established (register) by the country's clinical engineering society and the candidate has to attend a two years course plus a practical training in a certified hospital for several months.

There is no world standard or even basic requirements to be recognized as clinical engineer worldwide. Worst of all, there is no international entity that exclusively represents clinical engineering and where would be possible to discuss such matters, and perhaps reach a common sense regarding the recognition requirements.

All the recommendations regarding the responsibilities for clinical engineering are produced by isolated sources and strongly dependent on the countries' healthcare system requirements and state of the art. However, such isolated efforts can demonstrate a clear evolution not only on the increasing number of responsibilities but also on the way such responsibilities must be dealt with. Table 1.2 shows a set of the clinical engineering responsibilities listed by the CED/IFMBE in 1992 (Bravar, 2010), and the responsibilities listed by the ACCE in 2012 (Grimes, 2012).

Table 1.2 also shows an attempt to match similar responsibilities listed 20 years ago and listed in 2012. It can be noticed that not only the number of responsibilities had increased but also the way the clinical engineer must fulfill such responsibilities. While in 1992 the statements were very straightforward as the only responsibility derived from the clinical engineer, in 2012 the statements were to guide the clinical engineer to look for partners and collaboration among the health staff. For instance, while in 1992, the stated responsibility was "to provide technical consultancy on acquisitions and ensure the correct installation and testing of biomedical equipment", in 2012 the same item was to "Maintain effective communications with other stakeholders involved in the acquisition, use, and maintenance of healthcare technologies." Though they look the same in terms of duty, what changes is the way to deal with such duty. It is a more participative way to work with the several responsibilities this professional will have to deal.

As already mentioned, the number of responsibilities for the clinical engineer had also increased according to the evolution of the healthcare system. While in 1992, one of the responsibilities was "To develop software programs and hardware interfaces between biomedical instruments and hospital information systems," in 2012 the responsibility had changed to "Identifies and manages appropriate software upgrades, security patches and anti-virus installs for interconnected/integrated medical systems according to industry best practices." Such statements show that there happened a strong tendency for changing the word "Developments" to "Management."

In fact, depending on the complexity of the health unit, clinical engineers will not have time to develop software or hardware. Their task will be to find the suitable health

Table 1.2   Comparative Table Showing the Activities Established for Clinical Engineering by the Clinical Engineering Division/International Federation of Medical and Biological Engineering (CED/IFMBE) Defined in 1992 and the Ones Defined by the American College of Clinical Engineering (ACCE) in 2012

CED/IFMBE 1992	ACCE 2012
• To analyze the technologies available on the market	Report on the status of the healthcare technology program to management
	Report on industry developments, emerging technologies, regulatory changes, and other issues related to healthcare technologies
• To plan the replacement of obsolete equipment	Prepare and implement a healthcare technology (i.e., medical equipment) management plan
	Review the plan annually and modify as necessary to improve performance
• To provide technical consultancy on acquisitions and ensure the correct installation and testing of biomedical equipment	Maintain effective communications with other stakeholders involved in the acquisition, use, and maintenance of healthcare technologies
	Participate in standing and ad hoc committees (e.g., EOC, Safety, Capital Budget, QA, management) and Task Forces as needed
• To manage the maintenance of biomedical equipment and ensure its safety and effectiveness, making use of the hospital system's internal maintenance facilities and maintenance contracts drawn up with producers or service companies	Develop and implement clinical engineering department strategies, goals, objectives, and policies that are consistent with and support the organization's goals, objectives, and policies
• To prevent dangerous situations, through the acquisition of equipment compliant with national and international standards, and through the dissemination of information and international reports on defects in equipment available on the market	Maintain consistent standards of clinical engineering practice throughout the organization and ensure that all relevant regulatory, licensing the accreditation requirements (e.g., federal, state, and JCAHO) are met
• To lend direct support to medical staff using complex technology to perform clinical procedures and to coordinate the activities of health services support technicians modifying equipment or medical devices	Educate stakeholders on security and other implications associated with the proliferation of interconnected and integrated medical technologies
• To develop software programs and hardware interfaces between biomedical instruments and hospital information systems	Maintain current inventory of networked and integrated medical systems (including catalog of services, features, interconnections)
• To organize educational sessions on biomedical technologies for the Clinical Engineering, medical, paramedical, and administrative staff of health facilities	Define, develop, and manage the delivery of the organization's clinical engineering services … including consultative, technical, and educational services
• To determine optimal technological solutions for resolving clinical problems with the possible development of prototypes of equipment or medical devices, and to run clinical trials on any such prototypes prior to their industrial production	Identify, acquire, and manage resources (e.g., type and quantity of personnel, material, equipment, vendors) necessary to achieve clinical engineering's goals and objectives

technology (hardware or software) in the market and verify if it complies with the country's legislation and internal rules of the healthcare unit.

Due to the increasing complexity of the medical technologies during this past 20-years, new responsibilities to be accomplished by clinical engineers were added to the ones listed in Table 1.2 (Grimes, 2012). It can be noticed that many of these additional responsibilities have "integration" and "safety" as important issues. It may be alleged that some of the duties stated in the 2012 list are "embedded" within the ones listed in 2002. This may be so since the statements by earlier duties define "what" has to be done while the later mixed "what" and "how should be done." However, there are two important points here to call the attention: the inclusion of several new duties and the significant changes on the way to perform old and new duties.

— Identifies and manages appropriate software upgrades, security patches, and anti-virus installs for interconnected/integrated medical systems according to industry best practices
— Maintains current inventory of networked and integrated medical systems (including catalogue of services, features, interconnections)
— Works with stakeholders to insure effective deployment, integration, and support of new medical systems into legacy systems and non-medical elements of the organization's information infrastructure
— Identifies, acquires, and manages resources (e.g., type and quantity of personnel, material, equipment, vendors) necessary to achieve clinical engineering's goals and objectives
— Prepares and submits annual department budget and monitors expenses to control or justify variances associated with the approved budget
— Represents the organization's interests with respect to clinical engineering in the larger healthcare community through participation in professional organizations and industry initiatives
— Performs other related duties incidental to the work described herein
— Coordinates security management process including risk (e.g., criticality and probability) and vulnerability analysis and related documentation associated with interconnected/integrated medical systems
— Coordinates with stakeholders a process to prioritize, develop, and implement plan to manage/mitigate identified risks associated with interconnected/integrated medical systems by applying appropriate administrative, physical, and technical safeguards
— Maintains the integrity of the Food and Drug Administration (FDA) approval for interconnected/integrated medical systems
— Conducts investigation using risk management tools on incidents involving integrated medical systems and reports findings to appropriate stakeholders for follow-up action
— Monitors and adopts industry "Best Practices" to ensure integrity, availability, and confidentiality of data maintained and transmitted across interconnected and integrated medical systems
— Supervises clinical engineering and other staff as necessary in clinical systems integration and infrastructure support (e.g., hybrid reporting structure, project supervision)

Educational System for CEs

Though the list of responsibilities is well defined, the major problem is to prepare the clinical engineers to attend such responsibilities. Traditional professions (civil engineering, mechanical engineering, etc.) in several countries have a reasonable international harmonization and understanding about offering equivalent subjects according to the kind of knowledge needed to attend the responsibilities. Clinical engineering still faces the problem of different interpretations regarding the set of subjects to attend the same responsibilities. Apart from the initiatives made in events such as The Biomedical and Clinical Engineering Education, Accreditation, Training and Certification (BIOMEDEA) (Nigel, 2007), it was not possible to find in the literature any other attempt for international harmonization.

To worsen this problem, many training courses for clinical engineers are organized by biomedical engineering groups where most of the teachers are just biomedical engineering teachers and researchers. In some Biomedical Engineering groups, the teacher had never set foot in a hospital and, despite being specialized on medical equipment basic principles, know little about what is important to prepare clinical engineers to work in the health environment.

A survey done by this author on 16 universities, offering biomedical engineering courses, from Latin American countries, showed only six of them have programs to train clinical engineers and the kind of training can be quite different from each other (Table 1.3).

It also showed that the number of students for clinical engineering specialization courses is around 10 per year and drops to less than 10 per year for MSc, PhD, and others. Biomedical engineering under graduation courses has an average of 31 students per year. However, the survey showed that even the universities with under and graduation courses on clinical engineering do not offer specific subjects toward clinical engineering (Calil et al., 2015).

As for the certification system, the educational requirement to become a clinical engineer varies according to the country. It follows a sequence of the different requirements from country to country for a person to become a clinical engineer:

a) There is a need of a under graduation course on electrical engineering, the title of Master of Science in biomedical/clinical engineering, and practical training afterwards.
b) There is the need of a clinical engineering certification from a specialization course.

Table 1.3 Training Program on Clinical Engineering from the 16 Universities Consulted

Countries	Consulted Universities	Under Graduation	MSc	PhD	Especialización	Otro
Brazil	5				1	
Mexico	3	1	1		1	1
Colombia	3	1				2
Peru	1					1
Venezuela	1			1		
Argentina	2					1
Chile	1					1
TOTAL	**16**	**2**	**1**	**1**	**2**	**6**

c) There is only the need of a practical training on medical devices.

When there is the need of an under graduation course as requirement to become a clinical engineer (first option above):

a) The candidate has to have a background in electrical engineering.
b) The candidate has to have an under graduation course on technical subjects but not necessarily electrical engineering.
c) The candidate has to have an under graduation course, but no need to be on technical subjects. In this case, even architects can become a specialist in clinical engineering.

In some countries, there is an official regulatory organization that takes into consideration the background of the candidates. This means that one can be approved in a training course but can only have the responsibilities of a clinical engineer if the background is on electrical or sometimes mechanical engineering.

There is no general agreement about what are the professional backgrounds to become a clinical engineer. Neighboring countries from the same world region have different understanding as well as the requirements regarding his previous knowledge to practice his profession.

As already stated, in 2004, there was an initiative in the European Union to establish such requirements (Nigel, 2007) as well as the training subjects to become a clinical engineer. Unfortunately, this never was carried on and the diversity of understanding and requirements for clinical engineering remains the same as described here.

Clinical engineering societies also follow different models according to the country. While in the United States, France, Brazil, Italy, and other countries, there are societies dedicated exclusively to clinical engineers; in Germany they have a society that congregates biomedical and clinical engineering, bioengineering, and medical ICT. This may be a good solution for team integration within the healthcare system.

How Clinical Engineering Shall Evolve

As happened in the past years and in the present, clinical engineering will evolve differently as well as play different role in the healthcare area, according to the country advancement on the care for the patient. While some countries will discuss the need to keep a program of preventive maintenance due to the reliability and availability of complex medical devices (AAMI, 2015), others will still be arguing if the donor is complying with WHO's Guideline for Health Care Equipment Donation (WHO, 2000).

While some clinical engineers will deal or have already dealt with activities regarding Information Technology, Health Project Management, and Medical Equipment Usability, others will still be struggling to implement a reliable software for medical equipment maintenance management within the hospital.

Conclusion

In 2015, there is a great effort to organize the first ever International Congress of Clinical Engineering. Up to now, clinical engineers could only internationally interact to each other during biomedical engineering events. The discussions about the activities, objectives, advancements, requirements, and ways to harmonize the profession are confined within slots mixed with other subjects dedicated to biomedical engineering. Also, it is not rare that good articles generated by clinical engineers were not approved since the reviewers have a biomedical engineering background and know little about how important it is to clinical engineers to share practical experiences.

An international event exclusively dedicated to clinical engineering only shows that this profession is already sufficiently mature and evolved to discuss its own destiny but, most of all, to try to detach from biomedical engineering events. It does not mean here that clinical engineering should become separated from biomedical engineering but it is difficult to deny that issues involving both professional are quite different. In fact, the future trend for clinical engineering is more toward management of health technology in general, involving issues such as safety, finance, equipment assessment, interoperability, information technology, quality control, and education. To develop such issues it requires just a small knowledge of some subjects belonging today to the area of biomedical engineering such as biomedical instrumentation (including biomedical sensors) and biological signals. The remaining set of knowledge for clinical engineering comes from areas of economy, production engineering, health informatics and safety.

What was described in this chapter shows the huge gaps between the clinical engineering duties from one country to another. Such gaps not only involve the set of knowledge required for a person to be called clinical engineer but also the training methods, the certification systems, responsibilities, government recognition, legislation, and so on. It also shows that despite such gaps clinical engineers have as basic responsibility to care about healthcare technology. Some may have additional responsibilities and some may have research responsibilities.

Metaphorically speaking, clinical engineering worldwide seems like a rally race where you have pilots with several miles apart from each other but, eventually, all of them will have to drive through the same tracks and pass through the same difficulties. They will never cross the finish line as long as the healthcare technology keeps evolving. The race pilots will only augment or decrease the gaps among them. They can use different engines, improve the engines, improve the form of communication, use different kinds of tires but the tracks to follow will eventually present the same difficulties and lead to the same situation. As the society requires more and more safety, improves the technology, demands cost reduction, and establishes tougher rules, clinical engineering will have to evolve and follow the same tracks and difficulties that more advanced health systems as well as their clinical engineers had to go through.

References

AAMI, 2015. Roundtable discussion: Getting to the heart of the PM debate. Biomed. Instrum. Technol. 49:, 108−119.

ACCE, 2001. Enhancing patient safety: The role of clinical engineering, white paper, American College of Clinical Engineering, Plymouth Meeting, PA.

Betts, W., 1983. The changing role of the clinical engineer: The need to develop management skills, presented at the AAMI 18th Annual Meeting, Dallas, Texas.

Bravar, D., 2010. The evolution of clinical engineering and the development of digital and molecular medicine: Cultural and economic effects. European papers on the new welfare.

Bronzino, J.D., 2003. Clinical engineering: evolution of a discipline. Clinical Engineering. CRC press, pp. 4−12 (Chapter 1).

Caceres, C.A., 1981. Clinical engineering: A model development in medical personnel utilization. Med. Instrum. 15 (1), 8.

Calil, S.J., Velero, A.M., Novaes, L., 2015. The curricular structure and research lines in Biomedical Engineering Schools in Latin American − Submitted for publication.

Dolan, A.M., 2004. Clinical engineering overview. Standard Handbook of Biomedical Engineering and Design. McGraw-Hill, New York, pp. 3−16 (Chapter 36).

Dyro, J., 2004. History of Engineering and Technology in Healthcare Clinical Engineering Handbook. Elsevier, San Diego, California, USA, pp. 7−10 (Chapter 2).

Grimes, S., 2012. Clinical engineering domains and related roles. Reference Materials − American College of Clinical Engineering, http://www.ask.com/business-finance/many-hospitals-worldwide-a1b5559945db88af#full-answer.

Nigel, J.H., 2007. Biomedical and clinical engineering education, accreditation, training and certification - BIOMEDEA. IFMBE Proceedings, vol. 16. pp. 1118−1121. Medicon 2007.

Scott, R.N., 1976. Portrait of Clinical Engineering: The Report of a Study of the Role of the Professional Engineer in Health Care, CMBES Monograph 1976−1. CMBES, Ottawa, Canada, pp. vi−85.

World Health Organization, 2000. Guideline for Health Care Equipment Donation. WHO/ARA/97.3.

2

Clinical Engineering in Growing Markets: The China Case Study

Yadin David[1], Zhou Dan[2,3]

[1]BIOMEDICAL ENGINEERING CONSULTANTS, LLC; UNIVERSITY OF TEXAS SCHOOL OF PUBLIC HEALTH, HOUSTON, TX, USA [2]MEDICAL MANAGEMENT DEPARTMENT, PLA GENERAL HOSPITAL AND PROVOST, PLA MEDICAL COLLEGE, BEIJING, CHINA [3]CLINICAL ENGINEERING BRANCH, MEDICAL DOCTOR ASSOCIATION; CLINICAL ENGINEERING BRANCH, CHINESE SOCIETY OF BIOMEDICAL ENGINEERING; MEDICAL ENGINEERING BRANCH, CHINESE MEDICAL ASSOCIATION

Introduction

The world we are leaving in is continuously evolving where successful and sustainable world communities dependent on technology is at all-time high. Community life quality and health are of no exception. Healthier behavior and health-related services depend on technology for the delivery of its services. In particular the demand for and deployment of technology in the point-of-care, where services patients and care givers converge, is rapidly growing. This dependence, alongside with the limited availability of global resources, means that adequate strategy and methods for optimal management of the technology throughout its life cycle is critically important. The combination of engineering knowledge and management competencies is a pillar of the clinical engineering profession. Among the many evolving things in our world, perhaps the combination of the development of organized scientific knowledge, working in teams, and technical management is recognized as new competency, which is a common element of a profession including that of clinical engineering. The rapid change in technological literacy together with markets globalization suggests that creating and belonging to professional associations, especially in the context of healthcare technology rapidly growing systems, can be beneficial for individuals practicing in the field. The provision of information, advocacy, professional development, mentoring, and career opportunities is part of the offering of professional association, and clinical engineering and healthcare technology management are of no exception. The adoption of these elements is the reason that the Chinese Clinical Engineering association is rapidly growing in the size of its membership, in the scope of services it offers, and in representing their constituents to other associations and coworkers in the healthcare delivery industry. This chapter describes the evolution of clinical engineering in China and the recent activities of this association.

Clinical Engineering. DOI: http://dx.doi.org/10.1016/B978-0-12-803767-6.00002-7

Among the common element of every professional society around the world is the mission to service its community through the provision of the abovementioned activities including that of setting minimum level of demonstrating adequate professional practice competencies and conducting professional recognition program. In this regard, the Chinese Clinical Engineering Society (CCES) has to evolve with this vision in mind. The following segment in this chapter provides insight about and gives credit to the founders of the CCES for their endless efforts as they created and organized opportunities for clinical engineers in China to identify and pursue an improving competency in their level of practice and in their service to hospital effectiveness and to their communities. They are thus playing an increasingly important role as China pursues solutions to the challenges of expanding and improving the Chinese healthcare delivery system to better serve its people. These challenges include a system of extreme scale, cultural variations, current uneven distribution of resources, rapid economic growth that carries a desire for modernized healthcare, a corresponding large growth in the acquisition and deployment of medical technology, and struggles with payment models for hospitals services.

Overview of Clinical Engineering and Healthcare Technology Management Market

From innovation to retirement, safe and effective technology's life cycle requires collaboration and knowledge sharing between many stake holders. The stake holders include the inventors, manufacturers, regulators, distributors, installers, service personnel and maintainers, integrators, and users, to name a few. Regardless of the stake holder field of practice, competencies and skill structure all share in the commitment to "do no harm" and in the potential for conflicts of interest (Smith and Sfekas, 2013). Healthcare delivery in general and patients in particular expect that, like other professionals, clinical engineers be competent and ethical in their practice. Thus, when it comes to healthcare technology, clinical engineers serve as guardians of patient safety and quality care outcomes.

Life cycle healthcare technology management must integrate the following stages: Technology compliance, technology planning, technology assessment, acquisitioning support, installation services, commissioning, training, scheduled and on-demand service support, risk management, upgrades, replacement/retirement.

Healthcare technology has made improvements both in quality patient outcomes and in improving patient experience (Johnson, 2012). Healthcare technology management bridges the gap between theory and innovation and point-of-care practices where the service of clinical engineers is focused (Shemer et al., 2005). Clinical engineering program services aim at optimally managing the technology and to create sustainable environment of safe and efficient application of the technology in support of care services. The technology can vary from mostly mechanical device like patient scales to intensive intelligence-embedded equipment like robotic surgical system and various imaging modalities that range from ultrasound to Magnetic Resonance Imaging (MRI) and Positron Emission Tomography (PET) systems.

The market scope of healthcare technology management and servicing programs is mostly divided between Original Equipment Manufacturer (OEM) service organization, Third party or Independent service organization (ISO), and In-house service program (Clinical Engineering). Healthcare technology management and servicing of the technology can be provided by one of these service providers or in the alternatives as a combination of more than one of the providers. Variations in needs, resources, and location contribute to difference in program's terms such as scope of services, cost of service, response time, and quality. In selecting the provider, or providers, there should be systematic evaluation prior to initiating any implementation of these programs. Regardless of the program source, all of them depend on the competencies and skills of clinical engineering professionals to repeatedly and professionally delivering service excellence. The global growing in volume of healthcare services, and thus technology that is deployed at these care delivery institutions, creates global demand for educated and well-trained clinical engineering professionals. The U.S. Bureau of Labor Statistics predicts that highest percent growth (30% increase) of future jobs will be in the medical equipment servicing and ranked this job opportunity as the best during this decade (U.S. Bureau of Labor Statistics, 2015). The many phases of healthcare technology management program demand adequate engineering and management education that includes training also in communication, writing, and project management. Clinical engineers around the world who possess these skills are advantageously positioned for pursuing satisfying and critically important career.

The Government of China embarked on improving affordable and expanding access to healthcare services for its population through the National Healthcare Reform Plan. Additional USD123 billion has been invested with the goal of funding hospitals, make services in rural and urban areas equally available, and establish national medical devices system (Zhou and Ying, 2013). The anticipated growth of clinical engineering profession under such an environment can only be diminished by lack of clinical engineering expertise.

China, by 2020, expects to surpass the United States, will have the largest economy in the world, and by 2025, China's nominal GDP will hit USD38.563 trillion. This economy will be characterized by high consumption spending, strong currency rates, and favorable trade ties (Sinclair, 2009). China's healthcare sector is also developing at an astonishing rate. This remarkable growth is largely attributable to the country's increasing government spending, underpinned by robust economic growth, which has led to improved healthcare access and infrastructure, as well as the ongoing expansion of public insurance coverage and infrastructure for less developed parts of the country (Han, 2012). Beyond the growth in government spending, patients' ability to afford better medical care has increased (McKinsey & Company, 2014). Driving this investment is the country's rapidly increasing middle class and its aging population, which are both increasing the demand for healthcare. It behooves on global clinical engineers in the academia, government, hospitals, and industry areas that if they desire to reach full appreciation for such enormous reform plan impact, then they also be cognizant of influences in China, such as cultural ones, have on professional practices.

The Evolution of Clinical Engineering in China

A common element among most of the professional societies around the globe is the mission to serve the community through pursuant of continuing education opportunities, networking and knowledge sharing, and professional development that includes professional recognition and certification. In these regard, the Chinese Clinical Engineering Society known as the Medical Engineering Society (MES) of Chinese Medical Association (www.cceweb.cn) is no exception.

Following the 2012 World Congress of the International Federation of Medical and Biological Engineering (IFMBE) in Beijing, China, colleagues' interest in the state of clinical engineering in China have grown. Not many non-Chinese clinical engineers have had the opportunity to visit a Chinese hospital; therefore, a starting point can be the sharing of relevant facts about the health-care system there. In comparison to the United States, an important fact that contributes to the growth of clinical engineering is that there are about 10-fold more hospitals there. More than 60,000 hospitals in China do vary in size, complexity, designation, and affiliation. In the larger metropolises, hospitals with 2000- to 4000-patient beds are considered normal and national. National and local Chinese governments are expected to spend about ¥100 billion, or about USD15.5 billion over the next 7 years. This significant investment in the Chinese healthcare services magnifies the need for continuous growth and improvement in healthcare technology management expertise and thus the clinical engineering profession.

The roles and responsibilities of the hospital staff in these facilities are not an exact match to those that are common in the United States. For example, the purchasing process and purchasing decisions, and the accreditation of hospital operations, and the role of nursing staff are all different than in the United States. To become part of the care team, the MES chose to become affiliated with the professional Chinese Medical Association (CMA). Affiliation with the CMA was accomplished in 1993 and today the MES is one of the 82 academic branches of the CMA. The presence of clinical engineers is varied among hospitals, with the majority of hospitals still not having the benefits of having trained clinical engineers as their staff. The total number of clinical engineering and technical personnel, in China, is estimated at 100,000. The MES now has a membership of several thousands of individuals and they have achieved some influence in China's rapidly growing healthcare delivery system. The members of MES represent every region of this large country. They are practicing as clinical engineers as this profession is recognized elsewhere, namely, by providing technical support through tasks such as preparation of bids, acquisitioning of medical equipment and information technology, installation and commissioning, planning and functional development, quality assurance, devices utilization, service and maintenance, technical training, systems integration, and support of scientific research (David & Janke, 2004). Others practice as consulting experts in academia and teaching including administrative staff from educational programs and from manufacturing companies working in medical engineering research and development, production, distribution, and support. While organization structure of clinical engineering programs can vary between healthcare providers, a typical program organization in Chinese hospital may look like the chart in Figure 2.1.

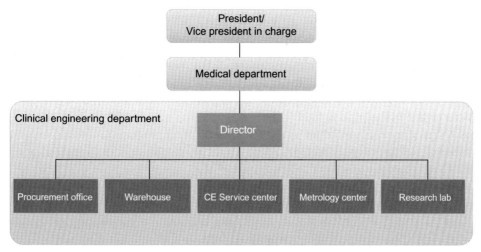

FIGURE 2.1 Common structure of CE program in Chinese hospital.

The MES has a strong liaison with government and medical sciences and technology associations, and has become an important society in improving healthcare delivered to the diverse and remote communities. Dr. Dan Zhou, who was elected chairman of the CMES in 2010, was instrumental in building the momentum for clinical engineering in China, for establishing its clinical engineering certification program, as well as serving as a co-opted member of the IFMBE Clinical Engineering Division.

One of the important activities that helps the MES is their ability to organize and deliver an annual conference and exhibition event that includes knowledge exchange, review of best practices, quality education and training sessions, and domestic and foreign speakers who are invited to contribute to the professional development of clinical engineering. After the development and adoption of policies and procedures for a national certification program in clinical engineering, the annual conferences for the last 4 years have also included training and the offering of the clinical engineering certification program. The program is based on a structure similar to the one practiced in the United States and includes both a written and oral exam. Consistent with the argument used in the United States and elsewhere, certification in China contributes to an overall improvement in competencies, recognition of those with demonstrated competence, and the standardization of clinical engineering practice in China. The program emphasizes life cycle technology management, hospital program management, engineering participles, regulatory compliance, risk management, patient safety, and integration of medical devices with information technology.

In 2011, the twelfth annual conference of the Chinese Medical Association & Clinical Engineering was held in Houngshan City. This conference cooperated with international organizations, and was supported by the Clinical Engineering Division of the IFMBE, by American College of Clinical Engineering (ACCE), and by entities such as the Emergency Care Research Institute (ECRI), and the German regulatory association. In addition, several foreign speakers came from a variety of countries. Over the last 3 years, each conference has drawn more than 1000 clinical engineers, biomedical equipment technicians, academicians, physicians, biomedical engineers, hospital IT managers, asset managers, healthcare IT managers, industry professionals, and administrators. Seven years ago, the MES established technical training and a system of certification examination for clinical engineers. More than 1000 people have now taken part in the training, and over 170 clinical engineers have passed the examination and have been certified. The passing ratio is about 60% of those qualified for and taken the examinations. This is the only established professional certification program of medical engineering in China, and they are actively working with the various government departments to set this program as a recognized national certification system.

The daily clinical engineering practice in Chinese hospitals includes carrying out a review and analysis of the quality control of medical equipment including technology performance. Between 2001 and 2009, their team of researchers completed a quantitative assessment of medical equipment and classification program of risk control, and established a quality system for medical equipment management and technical operations. At present, their effort and practice standards have been adopted in more than one hundred hospitals in China. In 2010, their team participated in the completion of a regulation of the safe clinical application of medical equipment, a regulation that has been issued by the State Ministry of Health. In some regards this level of direct influence in governmental actions exceeds that enjoyed by clinical engineers in the United States. As an additional indication of the professional evolution of clinical engineering in China, beginning in 2010 the establishment of the China Forum took place within RSNA meeting (The Radiological Society of North America) as part of their efforts to strengthen cooperation with the international community.

Like in the United States, the MES is actively promoting the convergence of IT and CE in hospitals, and medical engineering technology will play a more and more important role in the development of the digital hospital in China. As a result of rapid growth, China will have more new hospitals than the United States and many other countries, and it therefore has both the challenge and the opportunity to create systems from the beginning rather than retrofit both technology and contemporary good practice. Other projects include the classification and coding of medical supplies and the compilation of a basic training textbook for medical engineering personnel, which are also in progress. This year the Chinese association plans to carry out a nationwide survey and selection of the "Top Ten Service Engineers," and will organize a comparison of equipment warranty terms that purchasing groups are securing for medical institutions. Planning and coordinated organizations are working on the plans for the thirteenth Annual Conference of Chinese Medical Association (CMA) Medical Engineering Society (MES) scheduled for November in Ningbo, co-hosted by the Chinese Medical Association Medical Engineering Society and the Health Information Standardization Committee of the Health Ministry, with the support of Shanghai Medical Association, and Shanghai Clinical Engineering Branch.

The development of clinical engineering in China is part of that country's rapid economic and healthcare development, modernization, and resource and technology deployment (Bhuller, 2014). In support of their collective efforts, clinical engineers in China have become increasingly well organized and influential. This has included a large annual meeting and implementation of a continuing education program with focus on the promotion to achieving life-long professional competence and professional recognition through the awarding of plaques and certificates of achievement.

Professional Development of Chinese Clinical Engineers

Integral part of professional development is the ability of the practitioners to demonstrate compliance with recognized level of the required body-of-knowledge and with problem-solving challenges that are typical of clinical engineers practice. Professional development is recognized in China by examination program, consists of both written and oral segments, education, and experience qualifications, and success is recoded by the awarding of the clinical engineering certification.

Twenty years ago, China did not yet set up the occupation positions for CE in hospitals, and the clinical engineering staffs serving in the medical institutions lack the appropriate professional qualification certification and occupation access. Now, competent clinical engineering labor force is in serious insufficiency, while the number of medical devices are dramatically increased. Such situation causes serious imbalance. In large hospitals, the number of engineering technical personnel only account for no more than 5% of all healthcare staffs while the account is for 15−20% in developed countries such as the United States. The maintained labor force for medical equipment is less than one person per ￥10,000,000 ($160,000) on average. The clinical engineering technician with Bachelor degree or above only accounted for 37%, which is far lower than the level in developed countries. Labor resources of medical engineering in China cannot keep up with the development of new technology and equipment.

In order to promote the construction of clinical engineer training system to international standards, China began to explore the way to establish clinical engineer certification system 10 years ago. First, the international level clinical engineer certification was introduced. In 2005, Medical Engineering Society (MES) of Chinese Medical Association (CMA) hosted the first international level clinical engineer certification advanced training courses and initiated the certification examination program. With the expansion of the influence of certification examination, the Society totally held additional six sessions from 2005 to 2012. MES invited several international senior specialists to conduct lectures and exams, including professors such as Y. David, who is the former chairman and one of founders of the American Association of Clinical Engineering (ACCE), and Professors (emeritus) W. A. Hyman, J. Wear, and E. B. Sloane are all experienced leaders in clinical engineering and healthcare technology systems. The form and content of examination are in line with the general international standards with additional content specific for China. Following the verification of meeting the minimum educational and experience requirements, evaluation of the comprehensive knowledge capacity of candidate is taking place with the certification examination process consisting of two parts including written test and interview given in the English language.

All questions were written by the American clinical engineering experts based on their years of knowledge and experience and understanding of the practice in China. In the following six sessions of training workshops, more than 700 clinical engineering staff from hospitals, universities, or industry participated. Among them 219 passed both examinations and were awarded the level of clinical engineer certification. Of the successfully certified engineers, 90 are from the A level hospitals (main medical centers) representing 18 of the 32 provinces and municipalities throughout the whole China nation. Most of them are leading responsible positions of MESical engineering department and 37% even as senior engineers.

In the past 10 years, China always made efforts to establish its own professional clinical engineering certification system based on the lessons learned from other international clinical engineering certification programs. In October 2012, the process of clinical engineering qualification in China took a new important step forward. MES carried out the first Chinese Registered Clinical Engineer Certification training class and examination. Candidates were the junior engineering employee in large hospitals or new graduates who majored in MESical engineering. Throughout the country, it is estimated that there are more than 20,000 people who are qualified to participate in the examination. Registered clinical engineer certification is the basic admission examination to the occupational qualification of clinical engineering. The examination focuses on the basic theory and skills, and consists of theoretical exam and practical one-on-one interview and testing. China has established the Chinese exam question bank. So the theoretical exam questions for clinical engineer qualification can be randomly selected from that bank. The practical test requires candidates to present their plan for overcoming challenges in equipment evaluation, system integration, adverse event investigation, making repair versus replace decision, measurement, maintenance, and other pertinent topics. A professional committee made of clinical engineering experts selected from around the country evaluates the ability of every candidate and provides the final determination of the candidate competency. In 2012, 176 people enrolled the examination, 58 people passed the exams and obtained a registered clinical engineer certification. In contrast with the international level clinical engineer certification, registered clinical engineer certification covers a wide range of more junior medical engineering practitioners (Figure 2.2).

In the coming decade, China plans to set up two levels of certification system for international level clinical engineers and for registered engineers. The international level clinical engineer certification will mostly address the need for the senior engineer who will engage in clinical engineering for more than 10 years. In China, there are about 500 specialists who are qualified to participate in the examination, while more than 10,000 technologists are still expected to pursue registered engineers certification.

This system contributes its evolution of establishing continuous education system for practicing engineers and seeking programs to effectively improve (David et al., 2011, 2014) the quantity and quality of competent CE in China. In 2013, China established the Chinese Clinical Engineers Association (CCEA), which focuses specially on career development, vocational education, and representing the value of CE profession to society. Dr. D. Zhou, who was also one of the initiators of international level clinical engineers certification, was

FIGURE 2.2 Dr. Dan Zhou received the ACCE Antonio Hernandez International Clinical Engineering Award in September 2014 *(left to right, Y. David, J. Wear, D. Zhou, S. Wang, and B. Wang).*

selected as the first chairman of CCEA. In the next step, CCEA plans to recommend to the government to establish officially authorized national clinical engineer training and certification system which will accelerate the development of CE in China.

Perhaps one of the signs of the maturity of China clinical engineering association is its election to host the first international clinical engineering and healthcare technology congress in Hangzhou, China, in October 2015, under the umbrella of global clinical engineering program committee and the sponsorship of the IFMBE/clinical engineering division as well as of the many national CE societies from Italy, the United States, to Malaysia. The congress expects to attract multidiscipline professionals from many regions and healthcare sectors (www.icehtmc.com).

References

Bhuller, R., 2014. New Mega Trends in China — A 360 Degree Overview of Major Trends and Implications, Frost & Sullivan, February 18, 2014, Conference presentation, AusMedtech.

David, Y., Janke, E.G., 2004. Planning hospital medical technology management. IEEE Eng. Med. Biol. Mag., May/June, 73—79.

David, Y., Liu, S., Zhou, D., Li, B., Peng, M., Xia, H., 2014. Quality and safety must be the goal for hospital-based clinical engineering programs. China Med. Devices J. 29 (07).

David, Y., Zhou, D., Hyman, W., 2011. Clinical engineering development in China. J. Clin. Eng. 36 (2).

Han, D., 2012. Medical Devices & IVDs in the World's Second Largest Economy. Brandwood Biomedical. http://brandwoodbiomedical.com (accessed 10.05.15).

Johnson, C., 2012. Healthcare technology: A competitive advantage in improving the patient experience, Healthcare IT News, http://www.healthcareitnews.com/blog/healthcare-technology-competitive-advantage-improving-patient-experience (accessed 18.05.15).

McKinsey & Company, 2014. All you need to know about business in China. http://www.mckinsey.com/insights/winning_in_emerging_markets/all_you_need_to_know_about_business_in_china (accessed 15.05.15).

Shemer, J., Abadi-Korek, I., Seifan, A., 2005. Medical technology management: Bridging the gap between theory and practice. Isr. Med. Assoc. J. 7 (4), 211–215, http://www.ncbi.nlm.nih.gov/pubmed/15847198 (accessed 18.05.15).

Sinclair, J.A.C., 2009. China's Healthcare Reform, China Business Review, http://www.chinabusinessreview.com/chinas-healthcare-reform/ (accessed 10.05.15).

Smith, S.W., Sfekas, A., 2013. How much do physician-entrepreneurs contribute to new medical devices? Med. Care. 51 (5), www.lww-medicalcare.com.

U.S. Bureau of Labor Statistics, 2015. Medical Equipment Repairers, Occupational Outlook Handbook. http://www.bls.gov/ooh/installation-maintenance-and-repair/medical-equipment-repairers.htm (accessed 18.05.15).

Zhou, D., Ying, J., 2013. Development of clinical engineer certification in China. IFMBE News. 93, 25–27.

3

Clinical Engineering in Developing Countries

Mário Forjaz Secca

DEPARTMENT OF PHYSICS, FACULTY OF SCIENCE AND TECHNOLOGY, UNIVERSIDADE NOVA DE LISBOA, LISBOA, PORTUGAL

Introduction

Although clinical engineering is a very important and fairly well-defined speciality within biomedical engineering, its implementation in developing countries presents several specific challenges and difficulties. Here I will not go into technical details but, in a thought provoking way, I will approach some of the ideas, concepts, and difficulties that involve the implementation of clinical engineering in developing countries, with particular emphasis on Africa, based on my experience in Mozambique and with hospitals in Mozambique.

Talking about the problem of clinical engineering in developing countries is made more complex by the fact that there is a difficulty in defining what a developing country is. There are several definitions around the world, most of them based on economic factors; however, in the International Federation of Medical and Biological Engineering (IFMBE), we have been using the Human Development Index (HDI) (HDRO, 2014) as a way of separating developed from developing countries. HDI is a statistic incorporating indicators of life expectancy, education, and per capita income, and divides the ranking of the different countries into four groups of human development, each incorporating around 47 countries: very high human development, high human development, medium human development, and low human development. And here I will use the low human development as the group of developing countries being considered. Any index is imperfect and the adoption of this particular criterion can be argued, but our choice falls on this because, apart from the per capita income, it also incorporates a life expectancy indicator, which is closely related to the health situation of the country, bearing direct relevance to our theme.

Apart from the problem of defining what a developing country is, there are so many intrinsic differences between the various developing countries that a generalization of the implementation of clinical engineering in all these countries is almost impossible and very risky. Of the 42 countries in the lowest category of low human development, apart from five

Clinical Engineering. DOI: http://dx.doi.org/10.1016/B978-0-12-803767-6.00003-9

of them (2 in Asia, 2 in Oceania, and 1 in the Caribbean), all of the others are from Africa, a continent we tend to associate with the developing world, even though there are 4 African countries in the high human development group (Mauritius, Seychelles, Tunisia, and Algeria) and 12 African countries in the medium human development group. The fact that 37 of these low human development countries are from Africa seems to give the group some consistency; however, even African countries can vary extremely between them.

At the risk of oversimplifying, I will try to find some issues that are somewhat common in this developing country group and point out some that are more specific to Africa, or some parts of Africa. The aim is to defend that even in the low human development there are jobs and a necessity for clinical engineers.

Here I will not discuss the role of the clinical engineer overall, but rather point out some of the specific parts of the clinical engineer job or duties that apply or are particularly relevant to the developing countries.

Funding

The lack of funding overall and in particular in the health sector is the first and most important problem encountered, intensified by the bad management of those funds. As an example, in some countries, with an already impoverished economy, the annual budget for the presidency is higher than the combined annual budget of the ministries of Health, Education and Agriculture, which gives us an idea of what amounts are available to be invested in health, and the importance given to health. Also, in some countries the corruption involving the use and distribution of money is very strong, limiting even further the actual money that reaches the final destination intended. In this bleak economical environment, the necessity of hiring clinical engineers is obviously not considered a priority. But we should not despair, because some countries do have them and we have to build a case to push for more clinical engineers in the hospitals and health centers around the developing world.

Geography

Another difficulty present in developing countries is that there is a tendency to have a geographical imbalance toward the capital. This capital-centered approach means that the resources will tend go with priority to the administrative center, with diminished support and attention given to the smaller hospitals and health centers spread around the country. The difficulties in road access and communications just adds to this situation. It is possible to minimize this situation with an investment in a basic communication network, based on a growing mobile phone network, as is now available in developing countries. This should include access to central databases and teleradiology image exchange, among others. It would be important for a clinical engineer to be involved with the smaller hospitals and centers, which can start by helping to train local technical staff and interact with them at a distance. Making an effort to visit regularly the distant locations should also be a priority.

Healthcare Policies

Many developing countries have a distinct shortage of medical doctors, healthcare professionals or fully medically trained personnel and have adopted the concept of Primary Health Care (PHC) (WHO, 1978) for their health services around the country. This means that apart from the main hospitals in the major towns, the nationwide health services will be provided in "essential health care" units that allow universal health care to be accessible to all individuals and families in a community. One of the basic principles of the PHC unit philosophy is the use of appropriate technology. And here, although there is no need to have a clinical engineer in each unit, it is very important to have a clinical engineer in a particular region or district of a country, to take care of the medical technology available in all the units in the region. Since the medical technology in use should be accessible, affordable, feasible, and culturally acceptable to the community, it is important for the clinical engineer to know the community very well, be integrated with it, and also be creative and innovative, to be not only able to keep the existing equipment in good working condition and to supervise its proper use, but also to be able to create and produce new appropriate technology with the available resources.

Traditional Medicine

One of the difficulties encountered in Africa in particular is a cultural problem and relates to the prevalence of traditional medicine practiced by witch doctors. Added to that is the fact that the implementation of modern scientific medicine is very recent. We should bear in mind that this occurs not only in the countryside but also in the main towns and the capital. A lot of the population will go first to the witch doctor before going to the hospital or health center and they do not trust so much the hospitals and the medical staff, let alone all the modern technology that is presented to them. The clinical engineer should be very aware of this because many of these people will have a distrust of the medical equipment being used on them. An immersion in the local culture is advisable and his job should also include a diplomatic attitude of convincing the patient to accept the technology, without unnecessarily criticizing the traditional medicine the patient believes in (let us not forget that not all of traditional medicine is bad or lacks common sense).

Distinction Between BME and MP

In the developing countries, there is a lack of training programs and lack of specialized people, and the limited funds allocated for health will imply a scarcity of jobs available. In this setting, the distinction between biomedical engineers and medical physicists is not very rigid and there are several countries that do not separate the two job descriptions. A clinical engineer in a developing country setting should have a broad background, by training or by extra study, and be prepared to bridge the gaps between the basic sciences and the technology, covering a bit of the two fields to communicate with the medical staff from both perspectives.

Maintenance Difficulties

One of the consequences of lack of funds, exacerbated by a more immediate approach to the acquisition of equipments, is that equipment is often purchased without buying a planned and regular maintenance program, with the maintenance people from the manufacturers only being called upon when something goes seriously wrong with the equipment. This lack of regular maintenance results in the equipment having serious working and safety problems and unnecessarily long downtimes. Here the clinical engineer can have a very important role, first in convincing the administration of the necessity of maintenance programs and the cost-effectiveness of them, second in helping to keep a basic maintenance procedure where possible to prevent unnecessary damage, and third in serving as the interface between the hospital and the manufacturers to speed up the repairing process and keeping the quality control up to date.

This problem is further complicated by the lack of supplies and parts, either because of limited funds or no budget allocation, or because of difficulties in distribution and access. We should not forget that many of these countries have no manufacturer representatives or distributors locally and the parts will have to be ordered from abroad, with the corresponding custom difficulties, and be transported to the location. It will be required of the clinical engineer to demonstrate a lot of creativity and diplomacy to find solutions to keep the equipments working.

Diversity of Manufacturers

A lot of countries also suffer from a large diversity of manufacturers for a corresponding small number of equipments. A lack of a well-planned equipment acquisition plan, based mainly on immediate interests and availability, implies that the equivalent type of equipment in various hospitals and different parts of the country have different manufacturers. This will make it more difficult to have a concerted maintenance program with a resource saving single team. It is not in the interest of many manufacturers to go to remote areas to service one single unit of their catalogue, and the prices charged will be in general too high, so the clinical engineer should once again be versatile or train local engineers to service, maintain, and repair the equipment with the available resources, no matter how protective the manufacturers are about allowing other people to touch their machines. A stopped equipment can cost lives, which should be the main consideration for the decisions.

Donated Equipment

It is very common in the developing countries to compensate the lack of funding with the acceptance of donated equipment. However, very rarely is there a well-planned donation plan, raising specific problems with donations. These can be: a lack of specific spare parts; a lack of specific running supplies, like chart paper for electrocardiographs; manuals and displays coming in languages that are foreign to the receiving countries; and no contact with the manufacturers. All of this can make some of the equipment practically unusable, and it is not uncommon to see developing countries became graveyards of donated equipment from the richer countries. A proper donation plan and quality check of the donated material should be implemented to mitigate these consequences, and here the clinical engineer should have an active part in this dialogue with the administration and be involved in the management of donations.

Training

The medical equipment training programs in most developing countries are very limited, which means that there are few qualified clinical engineers or people trained to service, maintain, repair, and use equipment. These training programs should not concentrate only on clinical engineering but should cover both the engineer and the medical physicist sides, in a comprehensive and versatile way, and I believe that clinical engineers should be involved in the designing and planning of the programs since their experience and points of view are very important for a strong, consistent, and practical curriculum.

Conclusion

From the considerations discussed above, we can see that the clinical engineer in a developing country will encounter some specific problems related to his job and duties there. The clinical engineer will have to be aware of all these contingencies and limitations to perform the job to the full and will have to be very versatile in these settings.

Although a big challenge, a clinical engineer working in a developing country has a very important and rewarding job with an opportunity to leave a mark and help in the direction of a stronger and more coherent health service in the country.

References

Human Development Report. 2014. Sustaining Human Progress: Reducing Vulnerabilities and Building Resilience. HDRO (Human Development Report Office), United Nations Development Programme.

World Health Organization. 1978. Declaration of Alma-Ata. Adopted at the International Conference on Primary Health Care, Alma-Ata, USSR, 6–12 September 1978.

4

Clinical Engineering for Consolidated Markets: European Case Study

Heikki Teriö

CLINICAL ENGINEERING, KAROLINSKA UNIVERSITY HOSPITAL, STOCKHOLM, SWEDEN

Introduction

Consolidated market is a general economic term describing the development of a certain branch, where the companies grow through merging with each other. This gives certain advantages, for example, through large-scale production, better possibilities for investments, and possibility to centralized administration, and through these measures the companies wish to cut the operational costs and increase the profit. Europe and the industrialized world is very much a consolidated market. Actually, this type of changes can even be seen in the field of healthcare with the goal to achieve cost-effectiveness for the operations and to enhance the possibilities to utilize the latest medical and technological development. In consolidated healthcare market, the hospital managements are rather independent within the region or county organization, but the supporting services are often outsourced to production units serving several hospitals or even the whole region. This has happened in many of the western European countries and especially in the Nordic countries. This chapter will describe the situation in the Northern Europe and especially in Sweden.

The Healthcare Market

The healthcare structure and delivery systems in European countries are today different from country to country. There are differences in how the systems are financed and how they are organized. The differences depend not only on differences in recourses, economical, or material, but also on culture, demography, and geography. Several decades ago the differences were not as pronounced as they are today, since at that time the highly specialized healthcare was concentrated only to few university or military hospitals in larger cities in each country. In the countryside, there were few doctors who had sometimes very large

Clinical Engineering. DOI: http://dx.doi.org/10.1016/B978-0-12-803767-6.00004-0

geographical areas as working range. Today the number of doctors and nurses has increased and the average number of practicing doctors per 1000 population in Europe is 3.4 ranging from 2.2 to 6.2, according to the Organisation for Economic Co-operation and Development (OECD) statistics (OECD, 2014).

The basic financing of the healthcare system in Europe is through taxes and it is the dominating financing type in the Nordic countries. Most hospitals are owned and run by the society. However, some of the public hospitals are run as corporations, but the sole owner is the society: the region government or the County Council. There are some private companies running hospitals and especially primary healthcare centers, but as it is in Sweden these companies must sign a contract with the County Council where the parties will agree on how much healthcare services the County Council will buy from the company. On the other hand, the same thing applies even for the county-owned hospitals.

In 2012, European Union (EU) member states used in average 8.7% of their GDP (Gross domestic Production) to health spending. This is 1.4% more than in 2000, but it is 0.3% less that the peak of 9.0% in 2009 (OECD, 2014). All the Nordic countries spend more than the average in Europe, with Denmark on the top with 11.0% expenditure. The other countries use between 9.6% to 9.0% of their GDP. The Baltic countries had lower expenditures compared to the European average reaching 6.1% in average (OECD, 2014). These are facts that influence the investments on medical technology and what priorities the healthcare providers must do. However, the financing of the healthcare has always been discussed and there is never enough money. For example, back in 1983 the Swedish Medical Association had a discussion on what to do about the shrinking healthcare budgets and how it will influence the investments in new facilities and new technology.

Retrospectively, we can see that the different technological innovations have clearly changed the methods for diagnosis and therapy. The development of medical imaging systems is a good example of how technological development has changed and improved the diagnostics. The development from fluoroscopy to 128-channel CT or CT/PET has surely changed the way clinical engineers work together with medical staff and manufacturers. Today the clinical engineers at most Nordic hospitals support the radiologists when purchasing the systems, they work with installation of these systems together with the manufacturers' staff, and they are also responsible for the technical operation of these systems and how the systems communicate with other systems like the Radiological Information System (RIS) and the Picture Archive and Communications System (PACS).

Another example is the development of electronics; how several new functions have been added to life supporting systems or how development of semiconductors, new sensor technologies, and large-scale integration of components have made it possible to make the devices smaller and develop more advanced instruments, like robots for minimally invasive surgery. In addition, the development of nanotechnology has contributed a great deal to the development of artificial organs and new diagnostic and therapeutic methods.

Despite the arguments that medical technology is very expensive and that the investments on medical equipment are increasing the healthcare expenditures, the number of medical equipment has increased. The number of advanced imaging technologies can be used as an indicator of overall development of medical technology usage. For example, the number of CT scanners and MRI equipment has increased in all European countries. As shown in Table 4.1, which shows the OECD statistics between 2000 and 2011 (there is no data available from Sweden), the number of CT scanners has increased in Finland and Italy more than 50%, whereas in Hungary about 29%. The number of MRI equipment during the same period has increased more than 200% in Italy, more that 100% in Finland, and over 70% in Hungary.

The average number of CT scanners in Europe is 20.0 units per million population ranging from 7.7 in Hungary to 34.8 in Greece. Corresponding figures for MRI are 10.5 systems in average ranging from 2.8 in Hungary to 24.6 in Italy. There are, of course, several factors that have impact to this increase, but nevertheless it is an indication of the increase in usage of healthcare technology, and one can also establish that this increase is similar within other equipment groups used in healthcare. The increase of the usage of medical equipment shows how technology has become more and more important for delivery of healthcare and how its importance is growing steadily. Therefore, it is essential that the healthcare organizations have a medical equipment management program. This program has to ensure that medical equipment is appropriate to the clinical needs and that it functions effectively and safely.

The impact of demographic aging, caused by consistent low birth rates and higher life expectancy, is likely to be of major significance in the coming decades within the EU. For example, in Ireland, where the highest share of young people in the total population in 2012 was observed to be 21.6%, the proportion of the population aged over 65 years is expected to increase from 11% to 15% by 2021 (Layte et al., 2009). In Sweden, the most dramatic change will occur from the age of 80, where the number of persons will increase by 50% over the next 15 years. The share of older persons in the total population will increase significantly even in other European countries in the coming decades. This will lead to an increased burden on the society to provide for the healthcare expenditure required by the aging population. Many of the elderly persons will be relatively healthy a number of years, but there will also be a number of persons with multiple disorders. Many of these elderly persons know more about their disorders and they also know more about the possibilities that there is to cure or relieve the illnesses or disorders. These persons probably want to be treated at their homes or other places outside of the hospitals, which will need new technologies and new ways of nursing.

Authorities, national or international (e.g., European Union), and standardization organizations produce new or update old regulations and standards that impact healthcare delivery system and the technologies used. The healthcare providers must ensure compliance with the relevant regulations and standards, which requires understanding of existing laws and standards and keeping track on the changes in these. The regulations and standardization are developed in order to increase the patient safety.

The description above gives an idea of the market in which the organization dealing with the technology management in the healthcare has to adapt itself. This market is not static one, but it will change continuously. There will be new political requirements, new demands

Table 4.1 OECD Statistics over the Total Number of CT Scanners and MRI Units per Million Population in Some European Countries

Variable	Unit	Country	Year 2000	2001	2002	2003	2004	2005	2006	2007	2008	2009	2010	2011
Medical technology	CT, total	Per million population												
		Denmark	11.42	13.25	13.77	14.47	14.43	14.02	15.82	18.49	21.48	23.72	27.58	29.26
		Estonia	–	–	–	–	–	7.38	7.42	11.19	14.96	14.99	15.77	16.57
		Finland	13.52	13.69	13.27	14	14.15	14.68	14.81	16.45	–	20.42	21.07	21.34
		France	7.01	7.37	7.62	8.07	8.78	10.02	10.37	10.32	10.84	11.08	11.82	12.53
		Hungary	5.68	5.99	6.3	6.52	6.83	7.14	7.25	7.26	7.07	7.18	7.3	7.32
		Italy	21.13	23.01	24.05	23.92	26.23	27.82	29.29	30.55	30.96	31.85	32.17	32.62
		Poland	4.42	5.2	5.81	6.33	6.91	7.94	9.23	9.65	10.86	12.4	14.33	13.44
		Slovak Republic	–	–	–	9.12	10.24	11.35	12.28	13.77	13.76	13.37	14.1	15
		United Kingdom	5.35	6.88	7.29	6.91	7.02	7.45	7.53	–	7.26	–	7.52	8.07
	MRI, total	Denmark	5.43	–	8.56	9.09	10.18	–	–	–	–	15.39	–	–
		Estonia	–	–	–	–	–	2.21	3.71	5.22	8.23	7.49	8.26	9.79
		Finland	9.85	10.99	12.5	13.04	13.96	14.68	15.19	15.32	15.62	15.73	18.65	20.23
		France	1.65	1.83	2.4	3.17	3.85	4.78	5.19	5.48	6.06	6.43	6.96	7.51
		Hungary	1.76	1.96	2.26	2.57	2.57	2.58	2.58	2.78	2.79	2.79	3	3.01
		Italy	7.76	9.07	10.85	11.9	14.09	15.01	16.96	18.77	20.06	21.59	22.47	24.17
		Poland	–	–	0.94	1.02	1.91	2.02	1.94	2.7	2.94	3.7	4.69	4.77
		Slovak Republic	1.11	1.3	2.05	2.05	3.72	4.28	4.47	5.77	6.13	6.13	6.86	7.04
		United Kingdom	5.62	6.21	4.99	4.54	5	5.4	5.65	–	5.5	–	6.21	6.61

Source: Data extracted on 22 April 2015 12:37 UTC (GMT) from OECD. State: Estimate.

Dependences of healthcare development

FIGURE 4.1 Illustration how the society's development influences the medical and technical development and vice versa.

from the authorities, new medical and technical achievements, and of course changing economical prerequisites. Figure 4.1 illustrates how the changes in different elements of the society, politics, economy, and demography can influence both the development of medicine and the development of technologies.

Political decisions can create new national or international research concentrations on medicine and technology depending on, for example, increased demands from the citizens or because of the demographic changes. The economical situation at the time of decision can either facilitate the decision-making or prevent it. Also the achievements in medical and technical research and development in turn will influence the development of the society.

Clinical Engineering Work

The aim of the clinical engineering work is to provide the hospitals and other healthcare providers with engineering support, know-how and technology management, which is based on knowledge, competence, and experience, so that the patients can be diagnosed and treated cost-effectively in the best possible way. This aim should be commonly accepted and used in all clinical engineering operations worldwide. Those who make use of the clinical engineering in different forms expect high availability of the services and they also expect high reliability of the products and services they receive.

What is included in the clinical engineering work in different hospitals varies from hospital to hospital. Clinical engineers deal very often with issues that belong to other engineering fields within biology and medicine, like different sensors, materials, and development of technology for diagnostics and therapy. The American College of Clinical Engineering (ACCE) defines a clinical engineer as a professional who supports and advances patient care by applying engineering and management skills to healthcare technology (ACCE, 1992), which is a very wide concept. To advance patient care by applying engineering skills on the healthcare delivery system surely needs knowledge of all the subfields of biomedical engineering.

For more than five decades ago, the number of equipment used in the hospitals was not so large and some individual medical doctors had private contacts with engineers who helped them to construct and repair devices that they needed for their work. But, as time went by the number of equipment used increased and there was a need to start to develop more systematic management of the equipment. Therefore, for example, in Sweden the Swedish Society of Medicine and the Association of Graduate Engineers started a joint project in 1955 to investigate the impact of the medical equipment on the healthcare. This was the starting point for the more organized cooperation between engineers and medical doctors in Sweden. During the decades to follow, university and research clinics had their own engineers who constructed devices that were used in different studies. A good example of this is the development of the first pacemaker implanted in a human. This was a result of cooperation between the cardiac surgeon Åke Senning and engineer, but also a medical doctor Rune Elmqvist who worked together at the Karolinska Hospital in late 50s. Elmqvist constructed the pacemaker using commonly available components and testing the construction together with Senning on dogs, before the late evening on October 8, 1958, when one of the devices was implanted in a patient. This patient lived after the first operation for more than four decades and used 26 pacemakers.

The engineers at the research institutes helped also the clinics at the nearby hospitals outside their institutes when their equipment needed maintenance or repair. But, within a couple of years the other hospitals needed more support than they could get and they had to organize the management of the equipment differently.

Governance of the medical engineering operations in the beginning was usually left to the Clinical Engineering Departments (CEDs) at the hospitals. As the demands on healthcare have increased in respect of quality, economy, and safety, the development of the operations has become far from an internal matter exclusively for the CED, and therefore the hospital management today shows increasing interest to this development. The hospital management has a justified reason to question what clinical engineers are dealing with, how effective work is, and if it corresponds to healthcare needs. Corporatization and privatization give more reasons to this, since the ownership structure becomes clearer and it is necessary to make an inventory of the different assets.

The technical support was organized in the beginning in CEDs even at the regional and even many local hospitals. Overall technology management in hospitals was the primary application of clinical engineering and it had the focus on the safe and efficient use of medical devices so that the patients could be diagnosed and treated in a best possible way (Lamberti et al., 1997). This meant that clinical engineers ensured that appropriate, efficacious, and cost-effective equipments were available at the hospitals to meet the demands of quality in patient care. In Sweden, the importance of technology management have been well recognized and accepted since the hospital managements have come to the view that technology is an integral part of all major policy and planning decisions. Clinic vicinity and short response time are important to in-house clinical engineering and make up couple of their success factors. Therefore, the CEDs have been quite strong whereas the manufacturers' or vendors' own service organizations have always been relatively small in Sweden and in other Nordic countries. But, this is not the case in several other European countries, where vendors or very often third-party service organizations serve some specific device groups like infusion and syringe pumps or C-arches.

The CEDs in the Nordic countries have developed technology management programs in the sense Joseph Bronzino expresses it. "The development and implementation of a comprehensive technology management program requires a systematic approach. Such programs must embrace not only the technical aspects of maintaining medical equipment, but also the development of policies concerning equipment acquisition, acceptance, training, use, replacement and disposition. In essence, the primary goal of any technology management program is to ensure that the most cost-effective methods for the safe and effective operation of medical equipment are utilized" (Bronzino, 1992). During the last five decades, the character of the support has changed considerably from pure inspections and repairs of medical equipment to more managing and consultative work (Grimes, 2003).

The hospital managements have required the change because they need better knowledge of how the CED works and how the results can be measured. When these demands were stated, also a more national discussion about quality assurance (QA) had started and more or less all the CEDs started development of quality management systems. This adoption of a quality management system was a strategic decision by the CE management to meet the hospital management requirements. The different needs, particular objectives, the services provided, the processes the CED had identified, and the size and structure of the organization influenced the design and implementation of an organization's quality management system. The description of processes and documentation of known routines in the CED and at the hospital would help the responsible persons to fulfill their task and guarantee a safe and cost-effective use of medical devices in diagnoses, treatment, and care. The responsible person for all operations at the clinic is the head of the clinic according to the national regulations. The main processes described in the early stages could vary from hospital to hospital or county to county, but later there has been some harmonization. This harmonization was partly due to the Nordic cooperation group called NORDMEDTEK. The group was started in 1985 and its objective is to promote cooperation between the member hospitals and solve common problems. The member hospitals from the beginning were the university hospitals in the Nordic capitals, except in Sweden, where the Academic Hospital in Uppsala and the University Hospital in Lund were members. Today the membership has grown even to other university hospitals in the Nordic countries. The NORDMEDTEK group also published the Nordic Guidelines for Good Clinical Engineering Practice, which in its latest version from 2002 compiles with the ISO 9000-2000 standard and was distributed in the Nordic countries. The CEDs applied mainly ISO 9000 standard, but in some cases even ISO 13485 and ISO 17025 standards depending on the scope of their operations. In 90s, NORDMEDTEK was active in organizing pre-audits to help CEDs with their quality work. Today most of the hospitals have a quality management system, but far from all are certified.

NORDMEDTEK stated that the main goal was to design the processes to satisfy the demands from the main CED customers. The processes were divided into different activities designed to improve the medical services. These processes resemble the areas that Grimes mean that the future CE work should include (Grimes, 2010). The following six main processes were identified: (i) Strategic planning, (ii) Procurement, (iii) Device management and support in clinical use, (iv) Disposal of medical devices, (v) Knowledge management, and (vi) Research and development of medical devices. These processes would ensure safety and cost-effectiveness and were developed and defined through CED customers' stated or unstated needs. However, there are differences in which processes are used in practice. Very often only three processes are in use, but then processes 1 and 2, 3 and 4, and 5 and 6 are combined.

In the beginning of QA work, many CEDs in the Nordic countries had problems to identify their role in healthcare organization. The role that they traditionally have considered to have, as the hospital's own and only clinical engineering competence, was no longer obvious. The reason was the general tendency from the former subvention financed operations toward a more efficient needs and cost-controlled management of healthcare activities. In a fully developed consumer—seller system with competition and internal debiting, the clinical engineering organization does not hold any obvious and exceptional position vis-à-vis, for example, suppliers and external consultants. However, in some cases, the hospital management considered that some of the clinical engineering operations, for example, safety issues, still should be funded centrally and not be subject to customer dependency.

Many CEDs were not able to establish goals for their operations, because it was not clear what role they had, that is, what was the reason and why the hospital needed a support organization like CED. One reason for this was probably that even the clinics had no clear idea of what the CED could help them with, for example, to fulfill the authorities requirements on handling the medical equipment safely. But, a clinical engineering support organization can never take its own decision on the clinic's goals or quality level. Responsibility must be limited to what the clinic and the clinical engineering organization have agreed in agreements or job descriptions. Another reason for the indistinctness was that CED had, and often still has, difficulty to limit their activities to the core operations. The CE profile was thus unclear not only for their own staff but also for the "customers." One could compare the situation with, for example, departments of medical physics in the Swedish hospitals that in many ways have similar tasks as CED, but they have gained a much clearer profile. "Their" authority, the Swedish Radiation Safety Authority, has certainly helped Medical Physics in this development. CE does not have an equivalent authority, but they have to refer to different authorities and show their position as an in-house support for the clinics to fulfill the different requirements.

The main idea is to have an operation that is patient oriented, a statement that almost the whole medical world uses today. Also, it is important that the CED acts like one organization and uses same processes in spite of the physical location. The fact that healthcare is a consolidated market leads to dispersion of the operations to different physical location close to the main customers, which in turn leads to a challenge to act as one organization, but the unanimous way of CE operations creates prerequisite for the whole healthcare organization to operate in safe and cost-effective way. This means that CED will be responsible for the management of the medical equipment so that the clinic has access to the specified services, device functionality, and material at least in the extent they have agreed on and pursue an active safety work to prevent adverse events with the equipment caused by technical reasons.

It also means that the CED will facilitate education, research, innovation, and development in medical technology. The goals for the CE operations could then be formulated as to provide for actual equipment functionality, which is comparable with functionality of new equipment. CED must have close cooperation with the manufacturers and vendors and they must deliver services with a good quality. CED should also see that it will be an attractive employer in order to secure the future human recourses and that it is active in development of healthcare system it is working with.

The strategies to achieve these goals could then be, depending on what has been agreed in the contracts with the clinics, to support the patient flow through maintaining the clinic's medical equipment specified in the agreement. CED should also have a good control over the everyday work and a good dialogue with the clinics about the needs of development. It is also necessary to have a partnership with the vendors and an effective quality management system that will secure the unanimous way of working. For the hospital management, it is important that CED will contribute to build up an internal technical expertise that the hospital can take advantage of when developing new control systems for management of larger medical equipment entities (e.g., equipment in complete Intensive Care Units) and when investigating technical problems. CED must also develop the cooperation with universities and research institutes. With good leadership and teamwork, CED could achieve the provision of human recourses. CED should have individual goals for the work for every coworker and they should also have an individual competence development plan. All these strategies must harmonize with the core values of the hospital and be in line with the hospital's goals and strategies.

To describe the organization could be in its simplest way, a block diagram with the different sections and units that belong to the CED. But, when the hospital's core operations are organized in a thematic way, which is quite common nowadays, and if one would like to emphasize the process orientation with the clinical engineers working in the patient flow in the different themes, the presentation needs to be in a matrix type.

Clinical Engineering Processes

Strategic Planning

The latest technology always needs a large investment in both equipment and competence. Therefore, carefully prepared planning of technology investments can reduce the running costs and costs for maintenance most evidently. The planning includes cost−benefit analyses, return on investment, and life cycle cost analyses (Terio, 2010). These analyses are of high priority support for the hospital management which decides on investments. Technology management shall ensure that regulations are fulfilled and also includes activities focused on safety and efficacy of medical services. The strategic planning is not exclusively a process for CED, but is a process that is of utmost importance for the management of the healthcare provider, and will therefore include different professionals in multi-professional teams. An important factor in strategic planning is to secure that adequate competence for use of medical devices are available within the system. A successful investment means also that clinical engineers need to ensure an effective working relationship with all parties. For example, clinical engineers should serve as technical liaison between the hospital and the vendors, ensuring that everyone has all the information and access necessary for them to deliver equipment and services.

An important approach of the investments is to divide them to strategic investments, investments to keep the "level" up to date and investments to rationalize the healthcare processes. To plan for strategic investments, it is essential to do assessment that takes into account the national economy. In cities like Stockholm, it is essential to discuss what the different hospitals are suppose to do; what is their assignment since all the seven hospitals cannot do heart or liver transplants. The assessment of this kind will affect the future mission of some of the hospitals. However, this is in a system where the healthcare is controlled by political decisions. In any case, it is important to think of what is the future situation—3 to 8 years ahead. How are the medical methods developed, and especially how the technology is developed? Even in the case of investment, to rationalize a certain healthcare process, one must do economical assessment before the decision. The problem here is to calculate the effects of different alternatives.

There is a need of a very good system for economical follow-up. The calculation of the effects of the rationalization is very often subjective depending on the persons who are doing it. Strategic planning is part of the medical equipment management system that CED is involved in and essential to achieve good medical care of the patients. When the hospital plans to introduce new medical device and technology, the CED must actively support and take part in the procurement process. An important prerequisite for procurement processes is an interdisciplinary teamwork to ensure that all clinical and technical evaluations and other professional aspects needed are met in the medical technology assessment (Cram et al., 1997). This assessment shall clarify safety and regulatory matters and includes even different technical and clinical test in order to reveal the equipment performance. Procurement department is usually responsible to ensure that EU directives are fulfilled, that "Life Cycle Cost" analysis is conducted, and that the projects overall risk analysis is done. But even here CED has knowledge that is needed and their participation is necessary.

Acceptance test is an essential part of the procurement process and it can be conducted as purely technical and performed by the CED, or done in cooperation with the clinical users to ensure that the equipment is safe before being used. It shall also include registration in the hospital's inventory system for medical devices.

Device Management and Support in Clinical Use

Preventive maintenance and corrective maintenance including calibration of medical devices are basic activities in the process of device management. When the equipment is purchased, there is also a discussion of who is going to carry out the maintenance activities. In the Nordic countries, it is often a mixture between in-house and vendor services. The decision, who is doing the different maintenance tasks and to what extension, depends on the recourses and the costs. It is also possible to renegotiate the agreements when the situation has changed.

Technology management requires even an accurate inventory to keep track on the detailed information on the medical devices and systems. The objective of the inventory system is to record the detailed data produced from all activities of technology management and convert them into meaningful information. It should also be possible to monitor equipment performance, reliability, and cost-effectiveness, as well as to assist decision making in equipment acquisition and technology assessment. The inventory system should provide a comprehensive, expandable, and easy-to-use database of protocols for the performance of quality control, preventive and corrective maintenance, electrical safety, calibration, and acceptance tests of medical devices.

Very often CEDs in the Nordic countries developed their own systems, but they were vulnerable relying on a couple of individuals and very often poorly documented. When the demands on information quality and security increased, the CEDs were forced to evaluate a number of systems during the past decades, for example, HECS (Hospital Equipment Control System) developed by ECRI (Emergency Care Research Institute); MEMS (Medical Equipment Management System) developed under the BEAM European project; and BITMANS (Biomedical Technology Management System) developed by INBIT (Institute of Biomedical Technology, Patras, Greece). The systems that were used after evaluation were Maximo, which is an inventory and administrative system designed primarily for industry, QA-MAP, Norwegian system originally from the University Hospital of Trondheim, and MEDUSA, developed by Softpro in Sweden (Softpromedical, 2015). Today MEDUSA is the system that is most widely used in the Nordic countries. To address and satisfy particular needs that the different demands imposed by the resources, size, organization, and policies of the hospitals and/or the CEDs, the system is made modular and customized to a certain extent. Since CED has to perform continuous monitoring and evaluation of its performance in order to identify probable problems and measure its contribution to the quality of patient care, the system also monitors and measures a set of quality and cost indicators, allowing a continuous overview of the departments' performance in terms of productivity, effectiveness, and efficiency.

Clinical Engineering and ICT

The introduction of computers and Information and Communication Technology (ICT) in healthcare has had a significant impact on the whole healthcare delivery system. The integration of medical devices and IT has made it possible to use new technological solutions within healthcare. Medical information systems are today connected directly to medical devices in order to retrieve physiological data or to control the function of the medical devices. IT systems and software used in healthcare have obtained more crucial importance for the treatment and care of an individual patient. These systems are often even life supporting. Shortcomings and defects in the software can constitute a risk for injuries or harm the patient. Integration of medical devices and IT and the use of the devices in networks have made it necessary to study the integration of the two areas. The Swedish Society for Medical Engineering and Medical Physics (MTF) has conducted a project with the aim to improve patient safety through clarification of responsibilities for work with medical devices integrated with IT-products/systems that are used to collect physiological data for diagnosis and/or treatment of a patient and transfer this data through a network to a server/database (MTF, 2008). A new acronym, MIDS was proposed to be used for these systems that stands for Medical Information Data System. The use of MIDS increases the requirements for higher competence in the persons who handle these systems. It is also necessary to develop the cooperation between clinical engineers and IT engineers. The project resulted in a proposal for national requirements for competence that engineers working with MIDS should have and also guidelines for cooperation between clinical engineering and IT departments in order to fulfill the demands that directives and legislation state. Technical development and political decisions that direct the healthcare will require in the future that new interdisciplinary organizations will be created within CED and IT. In many of the Swedish Counties, IT and CE are already in close cooperation or belong to the same organization. Medical Informatics, CED, IT, and biotechnology will be integrated more and more in a common organization to support the development of healthcare. To ensure the positive development for increased patient safety, when using different computer systems, it is required that CED and IT organizations develop their QA-systems and work for certification of their operations. The two organizations can do the certification separately in the beginning but in the future they should do it together. Common product and system administration for CED and IT is recommended. By this way, the routines, processes, cooperation, and responsibilities will be ruled in a natural way. New functions like the System Integrator, described in the IEC 60601, 3rd edition, should be introduced as soon as possible in order to handle the MIDS in a proper way. This function should not be a physical person, but a group of professionals from CED, IT, and user's system administration.

Research, Development, and Education

Research

Research as it is at the university departments with PhD students is very rare at CEDs. The main reason for this is that CED does not have the right to examine the students. Therefore, most of the activities within R&D is to deal with development projects or support to research teams. These development projects include design, construction, production, and testing prototypes of medical devices. In recent years, clinical investigations according to the standard ISO 14155 have been included. Many projects have dealt with telemedical applications where CE has been responsible for the ICT-part whereas the clinical personnel have answered for the selection and contact with the patients.

The prototypes that are developed and that will be tested on patients in the hospital must be classified as in-house products. Also, there are a number of medical devices that are modified or used in a different way than the manufacturer has intended, even these products are classified as in-house products. Such in-house production shall comply with the European directives for medical devices according to the Nordic regulations and they must meet the same requirements for safety as the CE-marked products. CED has very often described processes for developing prototypes, risk management of the products, testing, and documentation. Therefore, CE is in lead of the project groups. The hospital managements have approved the systems for quality control of such in-house production.

In some hospitals, there is no in-house production and the research teams or doctors and nurses who have an idea to modify or develop something are directed to contact other departments or institutes. But, for example, in Sweden, all university hospitals have their own teams that support the innovation processes. The goal for these teams is to develop new, safer, and more effective forms of treatment by promoting cooperation between healthcare, academia, and industry. Especially, these teams address the medtech industry's need for an innovative center focusing on medical technology with direct access to clinical trials. At the Karolinska University Hospital, an innovation center, where the CED is very much involved, was created in 2011 to provide a unique environment in which the needs of healthcare can be addressed and will create a focal point for medical technology.

Education and Teaching

According to the regulations, the medical professionals must have sufficient competence and skills to ensure safe and efficient use of medical devices. The heads of the clinical departments have the responsibility to ensure that the department personnel have this sufficient competence in their clinical work. New technologies developed and new requirements for use of medical devices in new settings increase the requirement of competence and skills for the users. The healthcare personnel have to have the possibility for training that is included in the continuing education. Clinical engineers also plan for user training programs for the medical staff and education in basic subjects like electricity and gas technology in order to increase their awareness of risks when using medical equipment.

The human factor is the cause of, or contributes to, 90% of all accidents (Bogner, 1994). According to a report, 98,000 Americans die annually because of medical errors (Kohn et al., 1999). In this report, the use of sophisticated and complex medical devices is pointed out as a contributory cause of many of these errors. To enhance the human performance, the devices should be constructed according to the standards and automating processes should be used to remove the possibilities for human errors. Therefore, Karolinska University Hospital has their own Centre for Advance Medical Simulation, where clinical engineers work as instructors and with technical support (CAMST, 2015).

As a part of the QA system, CED maintains records on the relevant qualifications, training, skills, and experience of its personnel. In general, each individual shall on his/her own initiative make and forward proposals of the training they need. One goal for the individual coworkers is that they have the formal competence to be accepted as students for PhD studies at the university. There is a certification system in Sweden that the MTF has developed for more than 20 years ago. The requirements for certification according to the system are often used as a starting point for continuing education. The management of CED must make it also possible for the technical staff to study professional literature and visit trade fairs in order to get information of current developments in medical technology.

Teaching in technology management and different biomedical engineering subjects at the universities or in capacity building programs can also be a part of the tasks for a clinical engineer. The cooperation with the universities is very much depending on the distance to the hospital. In Stockholm the technical university is in the same campus as the Karolinska University Hospital, which makes it easy for the students to come to the CED for practical training or that the CE staff go to the university for lectures. Teaching on biomedical engineering subjects in both Bachelor and Masters level increases the possibilities for the CED to employ capable coworkers. It is also possible to get a number of smaller projects done as diploma work.

It is not only engineering students that come into contact with CED. Medical physicists, for example, can have lectures on ultrasound physics and its clinical applications, or medical and nursing students have lectures in bioelectricity or medical safety. Students in some special educations can also have advantage of the contacts with CED, for example, the students at the fellowship program, which is a cooperation between the Karolinska University Hospital, Karolinska Institute, and the Royal Institute of Technology. These students are in a group of four persons who usually are graduated medical doctor, engineer on master's level, economists and industrial designer. They get an introduction in clinical engineering in the beginning of their one-year education. During this year they work at some of the clinics at the hospital with a mission to study the workflow and to find out possible improvements. These improvements may lead to a need of a prototype and they can get help at CED with the production of it.

Capacity Building

Capacity building is one of the little more unusual assignments that CE can devote their time to. This is of course very much depending on that there are persons who have the competence and the interest to participate in such a project. Many European countries, the United States, and Japan and different donor organizations have invested in countries in Eastern Europe, Balkan, Central Asia, Africa, and South America. The investments have included medical facilities, and a significant amount of modern medical equipment has been introduced in the recipient countries. However, there has been very little support for installation, acceptance testing, maintenance, repair, documentation, and training. In the recipient countries, it is often difficult to find national competence that can meet all needs in medical technology. This is of course one problem for the governments when they try to provide modern healthcare to all citizens. There is a lack of clinical and biomedical engineers who can support healthcare technology management and development, which could give important contribution to the improvement of the healthcare system in their country. The university faculty programs in the recipient countries do not include specific education in clinical engineering. To ensure a long-term medical device management and maintenance function in the recipient countries, it is necessary to build such a competence base through capacity building. Therefore, engineering and medical institutions must be supported to provide courses in proper medical equipment use, safety, and maintenance. This education must also provide the managements in the hospitals with competence in device-related issues, physical infrastructure needs, investment planning, procurement, and training of medical staff.

In order to address some of these problems, Sweden has had a number of aid programs in different countries. Clinical engineering capacity building programs have been financed by Swedish International Development Agency (SIDA) and these programs have been part of large projects for strengthening the management and maintenance system for the healthcare services in recipient countries. The participating lecturers and supervisors from Sweden have been employed by CEDs in some of the university hospitals. Generally, the CE trainee program must include a broad spectrum of expert knowledge and the basics necessary for a professional occupation. The students in these programs should be able to use scientific results and problem solving concepts in practical applications. The programs were vocational and included both academic and practical modules. The programs were usually designed to include four semesters, where the academic modules lasted for 7−9 weeks and these modules were followed by practical training for 10−12 weeks. The theoretical subjects covered all type of devices, routines, and techniques used at hospitals and in the healthcare sector in general. The main focus was on the understanding of the application area, the functioning of the human body, instrumentation, management, organization, safety, and regulations. The practical part of the training was carried out at hospitals both in Sweden and in the recipient countries. The practical training at the hospitals was built on high participation in the day-to-day clinical work. The practical training included activities like inventory, repair work, planning of and carrying out preventive maintenance and safety checks, spare part specification, and providing user training. During the practical training modules, a mentor at the hospital as well as at the Swedish counterpart supported the students. Reports and assignments were handed out during the practical modules and the students filled in a logbook to make notes on progress and experiences. Follow-up seminars were organized during the practical modules and the students were asked to discuss and share their experiences from the practical training at such seminars. Also, a proposal for a curriculum for permanent education program and training of local lecturers within biomedical engineering was written and a plan for implementation was made in collaboration with local technical universities and health authorities.

Conclusions

Clinical engineering has become an important part of the modern healthcare delivery system. The clinical engineers in Nordic countries must cooperate with a number of other groups and teams in the hospitals. They are persons who can provide the decision makers with detailed information about the current status of the medical systems used. They also have an important role in the work for patient safety. To be able to accomplish this important task they have in the modern hospitals, the clinical engineers must develop their work and carefully watch how the healthcare delivery system is developing depending on the development of new technologies and the changes in the society.

References

Bogner, M.S., 1994. Medical devices and human error. In: Mouloua, M., Parasuraman, R. (Eds.), Human Performance in Automated Systems: Current Research and Trends. Lawrence Erlbaum, Hillsdale, NJ, pp. 64–67.

Bronzino, J.D. (Ed.), 1992. Management of Medical Technology: A Primer for Clinical Engineers, (Biomedical Engineering Series). Butterworth-Heinemann, ISBN- 10: 0750692529.

Cram, N., Groves, J., Foster, L., 1997. Technology assessment—A survey of the clinical engineer's role within the hospital. J. Clin. Eng. 22 (6).

Grimes, S.L. 2003. The Future of Clinical Engineering: The Challenge of Change. IEEE Engineering in Medicine and Biology Magazine, March/April.

ACCE, 1992. The American College of Clinical Engineering. <http://accenet.org/about/Pages/ClinicalEngineer.aspx>.

MTF, 2008. The Swedish Society for Medical Engineering and Physics. <http://mtf.nu/files/MIDS/MIDS_Rapport_Eng_2008_web.pdf>.

Kohn, L.T., Corrigan, J.M., and Donaldson, M.S. (Eds.). 1999. Committee on Quality Healthcare in America, To err is human: Building a safer health system, Institute of Medicine.

Lamberti, C., Panfili, A., Gnudi, G., Avanzolini, G., 1997. A new model to estimate the appropriate staff for a clinical engineering department. J. Clin. Eng. 22 (5).

Layte, R., Barry, M., Bennett, K., et al. 2009. Projecting the Impact of Demographic Change on the Demand for and Delivery of Healthcare in Ireland. ESRI Research Series 13 2009. Dublin: Economic and Social Research Institute.

OECD, 2014. Health at a Glance: Europe 2014. OECD Publishing, Access from http://dx.doi.org/10.1787/health_glance_eur-2014-en.

Terio, H., 2010. Procurement of medical equipment in Sweden. Zdrav. Vestn. 79, 156–163.

CAMST, 2015. Center for Advanced Medical Simulation and Training. <www.simulatorcentrum.se>.

Softpromedical, 2015. <www.softpromedical.se>.

5

Hospital Technology Management

Nicolas Pallikarakis, Zhivko Bliznakov

*BIOMEDICAL TECHNOLOGY UNIT, DEPARTMENT OF MEDICAL PHYSICS,
FACULTY OF MEDICINE, UNIVERSITY OF PATRAS, RIO - PATRAS, GREECE*

Introduction

The Universe of Medical Technology

The impressive progress in science and technology and more specifically in the field of biomedical engineering, over the past 50 years, has led to the development of new and innovative medical devices with a determinant influence on the healthcare sector. Healthcare delivery today is technology driven. Medical devices are used today in virtually every healthcare delivery process. According to a short definition adopted by the World Health Organization (WHO, 2011), "a medical device is an article, instrument, apparatus or machine that is used in the prevention, diagnosis or treatment of illness or disease, or for detecting, measuring, restoring, correcting or modifying the structure or function of the body for some health purpose. Typically, the purpose of a medical device is not achieved by pharmacological, immunological or metabolic means." As this definition suggests, many different types of products are properly classified and regulated as "medical devices" with an estimated number of over 1,000,000 products available on the global market, grouped in more than 10,000 medical device generic groups according to the two most established codification and nomenclature systems: the Universal Medical Device Nomenclature System (UMDNS) and the Global Medical Devices Nomenclature (GMDN). The products included range from syringes, wheelchairs, and hearing aids, to heart valves, implantable pacemakers, X-ray imaging, and laser surgical instruments. Medical devices (MDs) can be as simple as tongue depressors, stethoscopes, and thermometers or as "exotic" as PET-CT, cranial electro-stimulators, and robotic surgical systems. Apart from the MDs, medical technology today also includes *in vitro* diagnostics and e-health solutions, used to diagnose, monitor, and treat patients.

According to EUCOMED (EUCOMED, 2015), between 1980 and 2000, medical technology reduced hospital stays by more than 50% and between 2000 and 2008 by an additional average of around 13%, although in Europe, for instance, from an average of 10.4% of gross domestic product spent on healthcare only around 7.5% was attributed to medical technologies. Medical technology is by far the most innovative sector due to the high level R&D activities all over the world. This is demonstrated by the 10,000 patent applications filed with the

Clinical Engineering. DOI: http://dx.doi.org/10.1016/B978-0-12-803767-6.00005-2

European Patent Office in the field of medical technology in 2013, more than any other technical field. As a result of this trend to innovation, products have nowadays a typical lifecycle of only 2 years before an improved product becomes available. The global market of medical technology accounted for more than 400 billion dollars in 2013. With an estimated annual growth of about 5% over the next 5 years, it is expected to overpass half a trillion by the year 2018. All this universe of medical technology needs appropriate management in order for the benefits from its application to arrive straight to the patient.

Need for Management

The more diverse the medical devices are, the wider and more complex are the problems that can arise from their use. These problems include mechanical failure, faulty design, adverse effects of materials implanted in the body, improper maintenance/specifications, user error, compromised sterility, or electromagnetic interference among closely located devices. Whether equipment is used for diagnosis, monitoring of patient condition, or therapy, the healthcare facility should ensure that it performs as intended by the manufacturer. Furthermore, the widespread use of technology in medicine can also result in increased costs of delivering healthcare services. Therefore, there is a need to develop a proper infrastructure for evaluating, supporting, and managing biomedical technology. This led, some decades ago, to the creation and operation of Clinical Engineering Departments (CEDs), which are manned with experienced and qualified personnel can assume the responsibility for the safe, effective, and cost-efficient use of technology. Since the 1980s, these departments started using software tools as a necessary means for medical technology management. However, since that time, the medical technology environment itself has been extensively changing and the management tools have been continuously modified and updated in order to address the needs.

The relatively small number of equipment with easier to perform maintenance by in-house service has been replaced by tens of thousands of equipment per large hospital in the developed world, with very diverse, impressive, and expanding number of technologies, requiring special knowledge and skills to maintain. From the limited number of standards (starting from the IEC 601) and the diverse national regulations of the seventies, we passed to thousands of standards and a rather overregulated, partly harmonized, environment. From the limited availability of personal computers and embedded systems in medical devices, we have today computers everywhere, medical software applications considered as stand-alone medical devices and medical devices incorporating computers as a common denominator. The paper-based archiving and reporting systems and limited communication means gave place to new information-based management systems, quick and easy information access, and an overflow of data produced daily. At the same time, many interoperability and interference problems arose, which are now addressed by families of standards (HL7, DICOM, etc.). Manufacturer's or third party-based service provided to hospital is today a necessity in the majority of cases, whereas new approaches appeared concerning the use of equipment through leasing or under third-party contracts applying reimbursement by act. Although these developments led to changes in the management procedures and tools applied, their basic core remains essentially the same over the past three decades.

Medical Technology Life Cycle

The Overall Picture

The medical technology management in all of its aspects is a multidimensional optimization process, by which the efficacy, efficiency, cost, and risks of technology throughout its life cycle, as well as the replacement and investment strategies, must be accounted for. During its lifespan, a device passes through several general phases (Figure 5.1). It first starts with an innovative idea and then it passes through the design, development, testing, and evaluation stages.

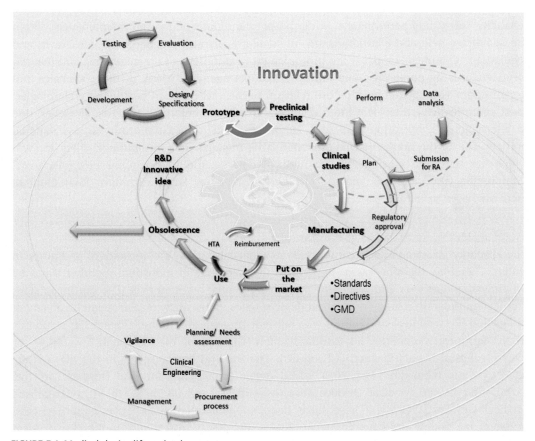

FIGURE 5.1 Medical device life cycle phases.

After approval, the device enters the manufacture phase, and then it is put on the market. Thereafter, the individual device starts its own real-life cycle, in many cases within a hospital environment and the effective management becomes necessity. That is where clinical engineers (CEs), as the key professionals, are assigned this critical task and that is where this chapter is focusing. It will, however, also address two essential areas that largely influence this work and where, at the same time, CEs play an important role: Regulations and Technology Assessment.

Regulatory Aspects

International Standards and Regulations

In order to protect public safety, while facilitating technological advances in the patient healthcare domain, international organizations and national governmental bodies have issued a number of standards and regulations, concerning the manufacture, marketing, use, and post-market surveillance of medical devices. As a result, the medical technology sector today is characterized by a high degree of regulation and standardization, aiming to ensure reliability, safety, and performance, under proper conditions of use and maintenance. Within the healthcare field, there are thousands of standards, clinical practice guidelines, laws, and regulations. Voluntary standards are promulgated by hundreds of organizations, and mandatory standards by numerous state and international agencies. Many of these agencies and organizations issue guidelines that are relevant to the vast range of healthcare technologies. Most countries have their own internal agency to set and enforce standards. However, due to international cooperation and trade, standards tend toward uniformity across national borders. Out of the many international standards organizations in existence, there are two which are the most concerned with medical devices: the International Electrotechnical Commission (IEC) and the International Organization for Standardization (ISO). Both of them are based in Europe.

Medical Devices Regulations in Europe

The disparity of standards became evident especially since the formation of European Union. Therefore, the aim was to harmonize the standards of individual member states by promulgating directives addressing the so-called essential requirements. The European standardization program relating to medical devices relies increasingly on European involvement in ISO and IEC technical committees with subsequent adoption of the ISO and IEC standards by the European Committee for Standardization (CEN) or by the European Committee for Electrotechnical Standardization (CENELEC). The medical device sector in the EU is regulated by three main directives: 90/385/EEC Active Implantable Medical Devices Directive (AIMD), 93/42/EEC Medical Devices Directive (MDD), and 98/79/EC In Vitro Diagnostic Medical Devices (IVDMD). These have been supplemented since by several necessary updates, due to new and emerging technologies (EC, 2007). In response to increasing safety concerns, legislative changes imposed stricter and more detailed monitoring and enforcement requirements for both Notified Bodies and national Competent Authorities in Europe. Additionally, the enforcement of a more rigorous but transparent new legislation in the form of two regulations is under elaboration and will overhaul the EU medical device directives. It is expected that a consensus could be reached in 2015 and these regulations will be voted in 2016 and will come into force after 2018. However, there are some concerns related to these developments, since the provision for additional clinical evidence requirements for high-risk devices will increase the regulatory burden on manufacturers, thus leading to delays and increased cost for innovative products to be placed on the market.

Medical Device Nomenclatures

An essential component of any MDs management program is the use of a coherent and comprehensive codification and nomenclature system. Two of the most adopted medical device nomenclatures used today are the Universal Medical Device Nomenclature System (UMDNS), created and maintained by ECRI since the 1970s, and, the more recently (late 1990s) European Union initiated Global Medical Devices Nomenclature (GMDN). UMDNS is a standard international nomenclature and computer coding system for medical devices (ECRI, 2015), officially adopted by many national systems and thousands of healthcare institutions, and for a certain period by the EU as well. GMDN is a joint CEN-ISO project, funded by the European Commission, and developed by harmonizing several existing regional systems in order to produce an international nomenclature. It covers medical devices and products as defined in the European directives, which are devices, systems, procedure packs (kits), accessories, and *in vitro* diagnostics. The GMDN (GMDN, 2015) is basically designed to cover medical devices as defined in the EU directives.

Unique Device Identification of MDs

Another system of major importance is the Unique Device Identification (UDI). It is actually a unique number pertaining to a medical device that enables the identification of different types of devices, and the access to useful and relevant information stored in a UDI database. Due to its specificity, the UDI can make traceability of devices more efficient, allow easier recall of devices, fight against counterfeiting, and improve patient safety. In the United States, in 2007, a Food and Drug Administration Amendments Act set out measures for implementing the UDI system. The UDI will not be a substitution, but an addition to the existing labeling requirements for medical devices. The European Commission leads an Ad Hoc Working Group at the Global Harmonization Task Force (GHTF) level in order to draft recommendations and guidelines to ensure that the UDI system will be globally applicable. The idea is to promote a global approach in order to avoid discrepancies between the different UDIs produced by the different members. The potential benefits of a UDI system include reduction in medical errors, improved adverse event reporting and post-market surveillance, and easier product recall process.

Technology Assessment

New technologies significantly improve clinical practice, but their rapid growth is making virtually impossible for care providers to keep pace with new advancements. Additionally, reluctance to change long-standing practices, as well as outdated education, restricts the uptake of new and potentially more efficient solutions. Growing concerns for more financial constrains, accountability, transparency, and legitimacy in decision-making processes imposed more evidence-based approaches. Health Technology Assessment (HTA) emerged exactly from the need to give answers and support decisions on the development, approval, and diffusion of health technologies. Although the scope of HTA is very large and the majority of reports address pharmaceuticals, the medical equipment sector has recently started gaining more attention. Its origin goes back actually in the 1970s, when healthcare institutions were under pressure to apply newly appeared costly medical equipment, especially in the diagnostic imaging area. A kind of precursor was the comparative evaluation trend of MDs that flourished in the 1980s, but was of limited scope and unable to cope with the very rapid changes of the markets. The growth and development of HTA and especially the hospital-based HTA, during the two last decades, focusing on medical equipment-related technologies, reflects the importance that this field is gaining. CEs being in the center of technology management, with a profound understanding of the technical issues, are well placed to play a pivotal role as members of the interdisciplinary teams working in HTA worldwide.

Hospital Technology Management and Clinical Engineering

As hospital technology continues to evolve, so does its impact on patient outcome, hospital operation, and financial resources. The ability to manage this evolution and its subsequent implications has become a major challenge for all healthcare organizations. Management of technology in the hospital environment starts with good matching between needs and capabilities. To be successful, it must integrate the technology procurement planning and management program at the hospital, with a final goal to address the needs of proper patient care (MHRA, 2014). This facilitates better equipment planning and utilization of the hospital's resources. CEDs started their establishment in the 1970s to fulfill this need. Today, the purpose of organizing and operating a CED in a healthcare facility is to obtain maximum benefit from medical equipment, while at the same time assuring its safe, effective, and efficient use (Dyro, 2004). These goals are achieved through the development of a comprehensive management and technical support program including services covering all aspects involved in the lifespan of every medical device. In general, the CED activities, related to the medical technology management, can be summarized as shown in Figure 5.2.

Implementing a comprehensive biomedical technology management program is a quite complex and multidimensional process, and depends on the skills and the background of the personnel involved. However, it contains in principle the tasks briefly outlined below.

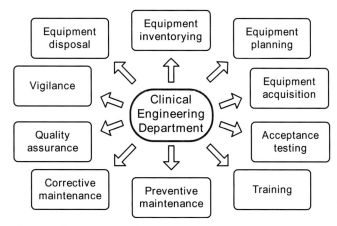

FIGURE 5.2 Clinical engineering department activities.

Medical Equipment Inventorying

The cornerstone of each program is the development of a complete inventory of the hospital's medical equipment. Inventorying consists of organizing records of all medical equipment items belonging to or used in the hospital. In this first step, correct identification of devices, where they are located, and who is responsible for their ownership and maintenance is essential. The CED establishes a method for recording all the information related to each equipment item. Records typically include data relevant to each device such as: inventory code assigned, device group, device type, manufacturer, model, serial number, location, and operating status. Additional information commonly recorded includes warranty expiration dates, purchase order and contract data, acquisition cost, installation date, and service vendor. Inventorying the existing equipment often turns to be a difficult task when it is not properly planned. A very critical factor is the use of a coherent and widely accepted nomenclature and codification systems, like the UMDNS or the GMDN. The future use of UDI will greatly facilitate correct MD grouping and easy tracing for vigilance purposes.

Medical Equipment Procurement Planning

Medical equipment procurement planning is the overall process of selecting equipment for a healthcare facility. Planning aims to ensure the availability of appropriate equipment at the right time in the right place. In addition, analyzing the market and choosing cost-effective technologies optimize the use of hospital capital resources and improve the quality of patient care. Medical equipment planning includes generally four major processes: needs analysis, technical evaluation, financial evaluation, and technology assessment.

Medical Equipment Acquisition

Decision-making process concerning plans for the replacement of outdated technology and the procurement of new medical equipment are also very critical ones, especially under conditions of limited resources. The CED is responsible for supporting the overall evaluation and prioritization of needs and assuring that new medical equipment will be harmonically incorporated in the existing technology and infrastructure of the healthcare facility. Medical equipment acquisition includes the following steps: definition of clinical requirements, assessment of environmental conditions, equipment market research, preparation of technical specifications, call for tender, offers evaluation, and contract signing. The whole process is a multidisciplinary team work, but CEs play a pivotal role. Correct technical specifications in particular are very critical in the procurement procedure, since they predefine the equipment that will be finally purchased. Recognizing the significant need of minimum specifications and requirements that should be considered before starting a process of purchase or donation of medical devices, WHO started the preparation of a first technical specification for medical equipment that provides guidelines in procurement and acquisition process to the interested parties [http://www.who.int/medical_devices/management_use/mde_tech_spec/en/].

Acceptance Testing

Acceptance tests are carried out upon reception of new devices or equipment. The purpose of acceptance testing is to verify that the correct devices have been delivered according to the contract conditions and technical specifications, and in some cases, whether the equipment is properly installed. In general, acceptance tests are intended to verify that the equipment received by the hospital performs its function and meets the manufacturer's specifications according to the requirements. An additional purpose, in the case of high-tech equipment requiring periodical quality control, is to obtain baseline measures that may be used as reference points to compare performance at a later stage and to resolve specific problems. Registering the specific values for well-defined tests as prescribed by the manufacturer, at the time of putting the equipment into service for the first time, allows CEs to check not only compliance with specifications, but also troubleshoot problems associated with the equipment during its whole life cycle. The acceptance test commonly includes: functional tests for compliance with manufacturer's specifications, visual/mechanical inspection, and QA and safety tests. CEDs must take great care that under no circumstances should a new not-inspected device be authorized for direct patient use. Having successfully passed all acceptance tests, the new medical equipment becomes part of the hospital inventory and is entered into the technology management activities of CED.

Training

Equipment-related education and training is considered to be a major priority for the CED, in order to properly apply the device according to its intended use and instructions of the manufacturer. It reduces adverse incidents due to user errors and minimizes risks to patients and hospital personnel, as well as helps to keep up-to-date with the technological changes and advancements. In addition, education of the technical personnel may result in optimization of departmental performance and productivity resulting in significant cost savings for the hospital. Therefore, training should address both medical and clinical engineering staff. Regular training of users is an essential element of any medical technology management program. Many studies reveal that the vast majority of equipment-related injuries are due to user error (Bliznakov et al., 2014). In order to significantly reduce these errors, clinical staff should be trained in the proper use of the medical equipment that they routinely operate. CED staff should also be trained in issues concerning maintenance, quality, and safety control of medical equipment. Vocational training seminars or courses delivered by manufacturers or third parties aim to improve knowledge and skills related to technical maintenance, and management issues, to ensure a safe and more efficient use of medical devices and to generally upgrade staff qualifications.

Preventive Maintenance

The main goal of a preventive maintenance (PM) program is to provide safe and effective use of medical equipment by keeping it in good functional condition. A well-scheduled PM program reduces the number of repairs required, lowers the downtime, and therefore reduces the inconvenience and frustration caused by malfunctioning or out of use equipment. The PM helps to eliminate hazards before they develop and/or become serious. There are certainly problems that occur suddenly and cannot be prevented by any PM program. However, many problems that occur because of equipment deterioration, caused by ordinary use, should be detected prior to leading to a malfunction. PM, according to the instructions provided by the manufacturers or best practice guidelines, undoubtedly improves operational conditions, increases reliability, prolongs the useful lifetime of the equipment, and prevents serious faults and adverse events.

Initially, the inspection schedule is set up based on manufacturer's recommendations or guidelines of third-party institutions. However, most schedules can be modified as experience is gained. If PM is performed too often, then valuable time and money are wasted, and if the inspection interval is too long, then PM is not effective. The basic units of the PM program are the periodic inspections. PM inspection basically consists of four different components: inspection and cleaning, change of predefined parts, functional testing, and safety testing, although not all of them need to be carried out during every periodical inspection.

Corrective Maintenance

Even the best and most effective PM program cannot eliminate the need for repairs, since it is impossible to prognose and prevent all equipment failures before they occur. Repair is defined as troubleshooting to isolate the cause of device malfunction and then as replacement or adjustment of components or subsystems to restore normal function, safety, performance, and reliability. Servicing and repairing the equipment is a significant part of CED activities and routine procedures should be established to make repairs as efficient as possible. They consist of performing the necessary actions for correcting medical equipment failures and re-establishing a normal operating status. Repairs can be accomplished either by the CED itself or by an outside vendor under warranty, contract conditions, or on call basis. In all cases, the responsibility of the CED is to ensure medical device repairs are properly carried out and documented. In some cases, failures are very simple, such as an inadequately adjusted control (operator error), loose connection, or wrongly powered equipment. An initial inspection by a CED technician can easily reveal such problems; the malfunction can be immediately corrected and the equipment is tested and ready to be put back into use. If the cause is not as simple, then a formal repair procedure should be scheduled. The decision for each repair depends on many factors, like: existence of warranty or contract coverage, device complexity, ability for the failure to be handled by the in-house CED, availability of spare parts, and appropriate technical documentation. In general, high-technology systems are often better handled by outside vendors because of their expertise and the availability of test equipment. In some cases, it is even obligatory to ask the manufacturer or its local representative, no matter how simple or complicated the failure is. All alternatives concerning repair should be well weighted by the CED, with a final goal to provide the most rapid and reliable repair at the lower cost.

Quality Assurance

Quality assurance comprises planning and performing quality control (QA) and safety tests of medical equipment to assure that functional and safety parameters are met according to the manufacturer's instructions and the corresponding standards. Quality control, depending on the equipment type, includes functional status inspection, QA tests, safety test (including electrical safety), and calibration. The QA protocols and their frequency of application should again be based on manufacturer's recommendations and internationally recognized standards and best practices. QA tests are often combined with the PM.

Risk Management and Vigilance

Risk management is closely related to both technology management and quality assurance programs. Each point of care involves risks not only for the patients, but also for the staff and other people visiting the hospital's facilities. The specific risks of each environment can be identified by conducting risk assessment studies, to identify them and put in place and to track and manage activities to reduce the potential for adverse events. Nowadays, medical technology has become much more reliable, efficient, and secure than ever before. Currently, all medical devices are produced and "placed on the market" according to international standards and have the necessary certification in compliance with the respective regulations worldwide. However, even the best-designed products, which are manufactured according to international quality standards, could potentially fail in clinical practice and cause serious problems or death to patients and staff. To avoid such adverse incidents, or more correctly to avoid their reoccurrence at a different time or place, the European Directives on MDs include provision for the establishment and operation of MD vigilance systems in the European Union (MEDDEV, 2013). These systems must be created under the responsibility of each member state and should collect reports of adverse incidents involving medical devices, perform investigation when appropriate, and disclose the information to all member states and the EU Commission, in order that necessary precautions are taken. It is basically a system for collecting and exchanging sensitive vigilance information. Any MDs vigilance system comprises all necessary actions taken by the users of medical equipment to inform the manufacturer and the responsible competent authority about an adverse incident. The general alerts management process falls under the responsibility of the CED and includes problem reporting to the authorities and application of root cause analysis (RCA) process where possible. On the other side, alerts acquisition and subsequent information sharing on incidents reported elsewhere; identification of potentially affected devices in use at the hospital; modification, replacement, or withdrawal of the equipment in question are even more critical. It is important to stress at this point that medical software is considered as an MD on its own. Studies performed on medical devices alerts revealed that almost half of the medical devices under recall were using software for their operation. Furthermore, two out of five medical devices using software have failed due to a problem in the software itself, indicating the growing use of software in medical equipment and its critical role in patient safety (Bliznakov et al., 2014). At the same time, there are efforts to create software systems that automatically identify publicly available adverse reports, alerts and recalls of MDs, compare them to MDs inventories, and generate targeted reports concerning devices in use in specific hospitals (Anastasiou et al., 2015).

Equipment Disposal

There are various reasons for the disposal of medical devices. These may include changes in the standards or new clinical procedures, possibly causing a device to become obsolete, even if it is still new in terms of years in use or functional aspects; technological changes in the performance criteria or accuracy that may dictate the device's replacement; or safety factors concerning increased risk or injury for patients or staff. Additionally, maintenance problems causing frequent and expensive repairs, excessive downtime, or nonavailability of repair parts. Alternative technologies with lower operational costs, or provided under more cost-effective leasing contracts, could be a reason for equipment replacement. The critical decision should be evidence-based mainly on data and information collected by the CEDs or found in available and appropriate HTA reports. Upon identification and decision of a device for disposal, the CED shall take certain actions. The device may be placed in storage and held as a back-up unit or disassembled in spare parts to support other similar equipment. When in good condition, it can be transferred to a research/training laboratory, or donated to a charitable organization or developing country. The last solution is to recycle the device or sell it as scrap.

Computerized Maintenance Management Systems

Management of hospital technology requires creation and maintenance of data files with all interventions performed on each individual equipment, covering its whole life cycle. These data must be easily retrievable and able to be presented to the end users in an appropriate way and converted into meaningful information. Since the 1980s, in order to satisfy the increased needs for medical equipment management, CEDs turned toward computerization as the only practical solution. Software tools initially developed "in house", specifically designed for the management of medical technology, started to be used to collect, store, and analyze data. These software applications offer many benefits compared to the manual paper work, including easy storage and retrieval of large amount of information, as well as data processing and analysis. These tools are known as Medical Equipment Management Systems (MEMS) or as Computerized Maintenance Management Systems (CMMS) (CMMS, 2011). There is an enormous progress of the ICT during the last three decades, offering new possibilities in data collection, storage, and retrieval, as well as improved user interface for the developers. Therefore, these applications have been radically transformed in terms of technologies used. However, the core of the essential information used remains practically the same. The use of a CMMS is to assist CEDs in performing their tasks, mentioned before, in a coherent, reliable, and easily documented way. Therefore, these systems shall cover all management activities of CEDs including:

- management of files for medical devices, manufactures, and suppliers;
- follow up of equipment acquisition procedures, from the request from the hospital departments to the acceptance tests of the devices;
- implementation and management of quality and safety protocols and procedures, including the necessary documentation and data, presented in an appropriate and comprehensible format;
- scheduling of all routine procedures, like preventive maintenance, quality, and safety inspections and tests;
- follow up of all corrective maintenance tasks;
- management and monitoring of training activities provided by the CED;
- monitoring of the overall performance of the department, using quality and cost indicators;
- easy access to and exchange of vigilance-related information;
- data analysis and report generation, either predefined or customized by the users.

In order to satisfy particular differences of available resources, size, organizational structures, and policies of the healthcare facilities, CMMS are usually implemented following a modular design approach, which allows setup customization. The ICT evolution has nowadays led CMMS to become complex database management tools, operating over network, interconnected with other hospital information systems and has imposed the need for continuous redesign and upgrade (Malataras et al., 2014). There are many CMMS commercially available on the market that provide most of the features needed. Screenshots of such a web-based CMMS are shown on Figure 5.3.

FIGURE 5.3 Screenshots of a web-based CMMS that offers a 24/7 access to the Medical Devices data, from any desktop, notebook, tablet, PC, or even a smart phone connected to the internet. (http://www.web-praxis.gr/).

However, several advanced CEDs still develop customized applications in order to better address their particular needs. The systematic operation of CMMS contributes considerably toward the improvement of effective medical technology management in healthcare facilities, with significant benefits relating to cost-effectiveness and safety. Moreover, CMMS facilitates information exchange between healthcare institutions and promotes the adoption of commonly accepted quality indicators and procedures.

Future Perspectives

Health technology will continue to grow and contribute to improvements in healthcare delivery. Apart from the new impressive technologies already announced or still at the R&D stage, probably the most important development that will influence the work and responsibilities of CEs will come again from the ICT side. Today's medical equipment either incorporates embedded computer systems or the equipment is computer-driven and increasingly interconnected or wirelessly connected to other systems. Therefore, a threat for these network-connected medical devices is to be infected or disabled by malware. This is also the case for m-health apps on smart-phones and tablets, and other wireless mobile devices used to access patient data, monitoring systems, and even patient implanted devices. The problems to deal with this situation is amplified by the varied responsibilities for purchase, installation, and maintenance of medical devices, with limited control over what is placed on the network and inconsistent training on the resulting risks and safety issues. On top of that, personal health devices, such as those used for home care or tele-health that are recently exploding in use, will also become interconnected to the Hospital Information Systems, thus constituting an additional factor of complexity. In the future, it is expected that largely interconnected and interoperable medical device systems, allowing automatic information flow to the right receiver, will proliferate. The e-Health and m-Health applications will rapidly expand, since more emphasis is expected to be given on elderly-related and chronic diseases that are seen as the main challenge of economic viability for the healthcare systems worldwide. These developments will also create huge data handling and ownership issues related to health technology, as well as privacy concerns. The abovementioned developments present also a challenge for CEs. It becomes clear that the strong need which appeared during the last two decades to manage medical device interoperability will become critical, since more and more medical devices are now connected to the IT infrastructure for data exchange (Sloane et al., 2014). Knowledge on how medical technologies integrate and interoperate between them and with other health ICT system networks becomes necessary. An important part of the needed knowledge and skills should therefore be directed to Health IT standards, software quality assurance aspects, including verification and validation and, in addition, human factors engineering and human machine interface, from both effectiveness and safety points of view.

In conclusion, the recent accelerated development of medical technology is expected to continue to reshape the healthcare sector. These technical trends are forcing CEDs to broaden their traditional role pertaining to the maintenance of medical equipment and to become involved in all aspects of hospital technology management and support, aiming to maximize the contribution and impact of their services on patient care. In this respect, a comprehensive hospital technology management program should also address this extended variety of technical and administrative issues and should encompass all necessary services related to safe and efficient operation.

References

Anastasiou, A., Bliznakov, Z., Deligiannakis, A., Giatrakos, N., Haritou, M., Ioannou, E., et al. 2015. Information extraction, matching and reporting for medical devices vigilance, In: 6th European Conference of the International Federation for Medical and Biological Engineering, IFMBE Proceedings, vol. 45. pp. 621–624.

Bliznakov, Z., Stavrianou, K., Pallikarakis, N. 2014. Medical Devices Recalls Analysis focusing on Software Failures during the last decade, XIII Mediterranean Conference on Medical and Biological Engineering and Computing 2013, IFMBE Proceedings, vol. 41, pp. 1174–1177.

Computerized Maintenance Management System. 2011. WHO Medical device technical series, http://whqlibdoc.who.int/publications/2011/9789241501415_eng.pdf.

Council of the European Communities. 2007. Directive of the European Parliament and of the Council of 5th September 2007 amending Council Directive 90/385/EEC on the approximation of the laws of the Member States relating to active implantable medical devices, Council Directive 93/42/EEC concerning medical devices and Directive 98/8/EC concerning the placing of biocidal products on the market. 2007/47/EC.

Dyro, J., 2004. Clinical Engineering Handbook. Academic Press.

Emergency Care Research Institute (ECRI). 2015. http://www.ecri.org, ECRI Universal medical device nomenclature system.

EUCOMED. 2015. http://www.eucomed.com.

Malataras, P., Bliznakov, Z., Pallikarakis, N., 2014. Re-engineering a medical devices management software system. The web approach. Int. J. Reliable Qual. E-Healthcare. 3 (1), 9−18.

MEDDEV. 2013. Guidelines on a Medical Device Vigilance System, MEDDEV 2.12-1 REV8. http://ec.europa.eu/health/medical-devices/files/meddev/2_12_1_ol_en.pdf.

MHRA. 2014. Managing medical devices guidance for healthcare and social services organisations. http://www.dhsspsni.gov.uk/dbni-2014-02.pdf.

Sloane, E., Welsh, J., Judd, T.2014. White Paper: New Opportunities for BME/CE Health IT Education. http://ceitcollaboration.org/docs/NewOpportunitiesBME_CE_Health_IT_Education.pdf.

The Global Medical Device Nomenclature (GMDN). 2015. https://www.gmdnagency.org/.

6

Management of Complex Clinical System

Elliot B. Sloane

RESEARCH FACULTY, VILLANOVA UNIVERSITY CENTER FOR ENTERPRISE EXCELLENCE IN TECHNOLOGY, VILLANOVA, PA, USA; EXECUTIVE DIRECTOR, CENTER FOR HEALTHCARE INFORMATION RESEARCH AND POLICY, OSPREY, FL, USA

Introduction

On September 8, 2015, the left General Electric jet engine on a British Airways Boeing 777 airliner departing from Las Vegas to England caught fire on take-off. The fire badly damaged the left wing, too, but the pilot and crew were able to abort the take-off and safely guide the jumbo jet to a place where all passengers and crew safely evacuated (British Airways Las Vegas plane fire: Eyewitnesses describe blaze).

Is this Boeing 777 incident relevant to Management of Complex Clinical Systems? In three words: "Yes, VERY relevant," because this incident illustrates how highly refined and effective the aviation industry's design, manufacturing, and maintenance have become. As we will discuss in this chapter, modern patient care systems have very similar profiles of complexity and risk. The clinical engineering profession needs to learn from, and adapt and incorporate, the same types of technology management systems that have allowed such safe, reliable, and cost-effective operation of the Boeing 777.

Actually, one of the most remarkable aspects of the British Airways 777 fire are that

1. After over 20 years in the air, such dangerous, extreme failures of Boeing 777 aircraft are extraordinarily rare, and
2. The only other documented Boeing 777 catastrophic accident involved an Asian Airlines in 2014, and has been primarily attributed human error (NTSB faults Asiana pilots for 777 crash, says complex automated controls a factor). The Asiana crash is actually a further testament to the robust design and maintenance of the Boeing 777; despite the near total destruction of Asiana Airlines' plane, out of 307 persons, there were 3 deaths and 49 injuries.

[For completeness, one other mysterious catastrophic 777 accident may ultimately be noteworthy: Malaysia Airlines Flight 370, which disappeared in 2014. As of the date of this writing, over a year later, only a single portion of a wing has been discovered, but the cause of the apparent destruction of the plane is not yet known.]

Clinical Engineering. DOI: http://dx.doi.org/10.1016/B978-0-12-803767-6.00006-4

Let us contrast the current state of the art of aviation technology management with the clinical engineering tools and approaches currently in use to manage ever more complex clinical technology systems in hospitals.

Similarities:

1. Both aviation and clinical engineering are responsible for managing with life-critical technologies;
2. Both are charged with supporting technologies many decades after original manufacture;
3. Both must manage complex systems, otherwise described as "systems of systems," aka (SoS). For example, each airplane is a composite of engine, navigation, braking systems, and information systems on the one hand, and each clinical technology system is a composite of multiple diagnostic, therapeutic, and information systems on the other hand;
4. Both must manage composite SoS made up of multiple vendors' products and systems, and some of the products and systems are incrementally added to or removed from the initial system long after initial manufacturing and configuration;
5. Both are under legal and/or contractual requirement to apply vendor updates and modifications to components and subsystems on an ongoing basis;
6. Both depend on a growing number of internal self-testing products and systems, removing user and manager from direct observation, control, and knowledge of each part of the system;
7. Both depend on a growing number of intelligent and artificially intelligent components and subsystems whose complete operation is not easily observed or modified by the user or manager;
8. Both depend on multiple disparate human-system interfaces, with risks that can only be mitigated with careful design, implementation, training, interpersonal communication, and resource coordination to ensure safe, reliable, and cost-effective operation.

At the same time, we must acknowledge there are many significant differences which prevent wholesale adoption of aviation's technology management approaches, including:

1. Each aircraft is designed, built, and managed by a single systems integrator such as Airbus, Boeing, Bombardier, or Embraer, whereas the individual hospital or integrated delivery network typically designs and integrates its own clinical technology systems;
2. Each aircraft design and application must be individually approved for commercial operations by regulatory authorities prior to sale;
3. Most aircraft component and subsystem which have life-critical implications are manufactured, maintained, and repaired under regulatory control for quality and safety;
4. Operation and repair of most aircraft components may only be legally performed by trained and licensed personnel;
5. The range of non-human threats to an airplane and passengers is finite, and generally limited to a very well-characterized—and testable—range of environmental factors, so airplanes are designed to withstand or perform within known envelopes of risk, and that

is quite unlike many still poorly understood diseases, widely variant human anatomy and genetics, and the side effects of medicines and other therapies which intermingle in more variations than science currently understands;

6. Unlike the airplane environment, the human disease environment is capable of aggressively evolving to kill each patient;

7. Airplanes are not evolving to meet individual requirements, BUT clinical technologies are evolving to deliver "precision medicine," which may not only be custom configured to individual needs, but may persist near, on, or inside humans for their entire life;

8. The wear and tear on airplanes is finite; airplanes—and pilots—are generally no longer in commercial service beyond their 6th or 7th decade, but there is a huge and growing population of humans with life spans exceeding 8, 9, and 10 decades, and eHealth and mHealth research is delivering a growing number of wearable and implantable technologies every year!

Foundational Principles and Concepts

In order to manage something, one need an appropriate way to identify the "thing" and adequate ways to measure and display its input and output. The process of "managing" implies some process of intervention to improve or optimize the output to meet a desired goal or target.

In this chapter, we are addressing "clinical systems," which we can coarsely describe as a combination of technologies, processes, and human actors working, presumably, to deliver and improve human healthcare. A "simple" clinical system example is a physician taking a patient's pulse by feeling the blood beating in an artery in a patient's neck or wrist. If one adds in inexpensive pulse oximeter device, available for just a few dollars online from eBay vendors, we can observe a more sophisticated, but still simple, clinical system that can provide oxygen saturation and pulse rate to aid the physician's diagnosis and therapy. Combining the physiologic data with the physician's observations, the patient's explanations, and the context of the patient encounter often allows the physician to make a reasonable diagnosis and begin a suitable therapy.

For the clinical engineer, "management" of relatively simple clinical systems like the one above was straight-forward. In the fourth issue of its first 1971 year of publication, The Emergency Care Research Institute (now ECRI Institute) published a Health Devices Journal issue describing their now-archetypal Equipment Control Program. As chronicled by the US Department of Commerce's treatise on Acquisition and Maintenance of Medical Equipment, the ECRI Equipment Control System became a global hospital staple (The Acquisition and Maintenance of Medical Equipment).

ECRI's early medical device management framework laid out an orderly process of planning, testing, and documenting the safe and accurate performance of medical devices. The process was based on ECRI's founder, Dr. Joel J. Nobel's prior experiences and observations as a small aircraft pilot and former medical officer on a US Navy nuclear submarine.

ECRI published the first testing protocol for defibrillators the month after its Equipment Control issue in its fifth 1971 issue, and, over the decades released widely adopted and translated Inspection and Preventive Maintenance Manuals and Checklists (IPM, ongoing), Hospital Equipment Control System (HECS, 1985) software, and Life Cycle Financial Analysis tools (Devices and Dollars, 1988).

The pre-purchase planning, procurement, pre-patient-use testing, and ongoing maintenance protocols were all designed for two goals: assure basic patient safety and ensure the physician and nurse were able to rely on correct measurements to guide their decisions.

The pre-purchase analysis of medical devices and systems is still very important, but it has become such a complex field that a separate chapter—or book—on micro- and macro- "health technology assessments" and "comparative effectiveness assessment" are needed. However, the clinical engineer must play an important leadership role in all such pre-purchase analyses, because he or she often has the most detailed understanding of the life cycle patient safety, regulatory compliance, system integration, and maintenance support issues for each product being considered. In fact, since the publication of ECRI's Devices and Dollars in 1988, it has become abundantly clear that poor pre-purchase planning can lead to catastrophic consequences.

Competent pre-purchase planning and analysis is critical for a second reason: no amount of post purchase testing or repair can build quality back into an inferior product. The aviation industry has learned this lesson well, and by using techniques like ISO 9000 quality practices, pre-purchase product and vendor qualification has become a bedrock principle.

Managing the Evolution to Today's Complex Clinical Systems

Many simple electromechanical test devices, like electrical meters and pneumatic gauges, allowed the clinical engineer (CE), and later, the biomedical equipment technician (BMET) to test and adjust or calibrate each medical device, allowing management of ever-growing inventories of electromedical devices.

Between 1970 and 2000, the strategies for managing clinical technologies went through several refinements to the core ECRI strategy. Most notably, Fennigkoh contributed the refined concept of "risk-assessment" to medical device maintenance interval determination. He demonstrated a method for computing a risk factor for each medical device in a hospital's inventory, using attributes like frequency of failure, patient safety consequences of failure, and operational consequences of failure to segregate high-risk products like defibrillators, infusion pumps, or ventilators from lower risk products like oxygen flowmeters or suction pumps. This segregation ultimately allowed hospitals to adjust IPM intervals from an initial annual or biannual cycle to coarser IPM intervals. In other words, by using a risk classification method, higher risk products are inspected more frequently, and lower risk products less frequently.

In 2001, Wang et al. demonstrated that the ECRI and Fennigkoh frameworks could be effectively adapted to manage the safety and performance of a very large fleet of medical devices, regardless of device age or service demands (Wang et al., 2001).

By 2009, Ridgway began adapting the Reliability Centered Maintenance (RCM) concepts from the aviation and power industries to health technology management (Ridgway et al., 2009). The RCM approach is data-driven, using past maintenance statistics to improve current maintenance processes. Some maintenance activities, like electrical safety testing of many double-insulated battery-powered devices, contribute no measurable benefit to product reliability. Such tasks can therefore be reasonably eliminated from the clinical engineer's IPM program, or they can be sampled on an infrequent basis to ensure that significant deterioration or risk is not imminent. Other maintenance activities, like tearing down and rebuilding an aged ventilator (or jet engine), can predictably cause costly—or unrepairable—damage to otherwise sound components. In such cases, a better strategy may be to put the ventilator on a more frequent inspection cycle and begin planning for replacement.

Boeing and GE used the RCM approach to design the 777 for extremely reliable service. The GE engine, for example, was designed with numerous internal sensors to monitor the "health" of the engine. Instead of aggressive engine rebuild cycles, GE and Boeing increasingly relied on sensor data to indicate deterioration. In addition, GE and Boeing developed methods of deploying ongoing field maintenance, updates, and modular replacement that could be performed in the field, between flights, to minimize complex and lengthy service activities that could take an airplane out of service for weeks at a time (Boeing Improves 777 Maintenance Program).

In aviation, the process of RCM to improve safety and reliability never ends, and new information allows Boeing, for example, to constantly improve its maintenance techniques. For example, Boeing identified that serious maintenance mistakes were often repeated by the same technician. For example, once a technician installed an oil filter incorrectly, it was likely the same mistake could be repeated. If several engines on an airplane had defective oil filter installations simultaneously by the same technician, the possibility of multiple simultaneous engine failures could have catastrophic results (Multi-Engine Maintenance). Boeing therefore changed its recommended maintenance strategies to avoid the same technician performing critical service to multiple engines on the same aircraft at the same time! (ETOPS-Based Maintenance Guidance). This very simple, nearly "free" alternate maintenance strategy has a high yield: it prevents a single human error from disabling the redundant and resilient systems that Boeing designed into their commercial jet aircraft.

The Rapidly Evolving Complex 21st-Century Clinical System

For the second half of the 20th century, most medical devices were designed for stand-alone operation. In fact, until integrated circuits and batteries were integrated into medical device designs in the mid- to late-1970s, most medical devices were also single-purpose. By the 1980s, however, combination devices like "battery powered defibrillator/monitors" began to appear. Such devices were necessarily single-vendor products, that is a company like Physio Control or American Optical had the system integration responsibility, much like the aircraft industry (Defibrillator: AO Pulsar 4).

By the early 2000 period, two key factors changed the status quo:

1. Olansen and Rosow demonstrated that medical devices could be designed, emulated, and changed using commonly available software, microprocessors, and sensors (Olansen and Rosow, 2001);
2. Rosow et al. demonstrated that clinical technologies could be successfully and meaningfully integrated with hospital information systems and infrastructure to provide improved executive and clinical decision support (Rosow et al., 2003).

The effect of these two breakthrough activities is still playing out to this day. In 2004, the IHE Patient Care Device Domain began formal development of international medical device and health information system integration standards (IHE Patient Care Devices), which, in turn, led to several ANSI-approved HITSP documents that were funded by the US Department of Health and Human Services through 2010 on Remote Patient Monitoring (Interoperability Specification) and on General Device Connectivity (Technical Note).

Today, medical device integration with hospital information systems is one of the most active areas of innovation in clinical engineering. Numerous regulatory and research summit meetings have produced large volumes of testimony, and white papers, including:

1. FDA Medical Device Data System guidance (Medical Device Data Systems, Medical Image Storage Devices, and Medical Image Communications Devices)
2. FDA Unique Device Identification regulation (Unique Device Identification—UDI)
3. FDA Medical Device Software Validation guidance (General Principles of Software Validation; Final Guidance for Industry and FDA Staff)
4. AAMI/FDA Wireless Medical Device Workshop (Healthcare technology in a wireless world)
5. AAMI FAQs on Wireless in Healthcare (FAQs for the wireless challenge in healthcare)
6. AAMI Wireless in Healthcare Resource Site (Wireless technology)
7. AAMI/FDA Summit on Medical Device Interoperability (Medical device interoperability)
8. FDA/FCC Joint Meeting on Wireless in Medical Technology (FCC/FDA Joint Meeting on Life Saving Wireless Medical Technology Day-1)

Currently, a working definition of Complex Clinical System includes the following characteristics:

1. Reliance one or more ICT technologies (microprocessor, programs, and communication);
2. ICT-based clinical calculations and/or interventions for multiple clinical parameters; and
3. User-interface that allows physician or patient intervention or control;

In addition, a Complex Clinical System may optionally include:

1. Device-to-device data sharing (e.g., morphine infusion pumps that can receive pulse oximetry or respiration monitor data and alerts);
2. Memory that allows storing and recovering data patterns and trends (e.g., smart eye surgery devices that allow superimposing prior corneal scans from archived images);
3. Computational tools for automation and/or to enhance predictive power and/or user-decision support (e.g., smartphone apps that take reported food consumption, exercise, insulin dosing, and glucose monitor results into account);
4. Automated patient or operator safety and risk detection, alerting, or intervention (e.g., smart infusion pump system that alarms or halts dosing when a morphine overdose calculation exceeds critical thresholds);
5. Internal or external software that enables self-testing, self-trouble-shooting, self calibration, and/or self-repair (e.g., power-on self-testing, or recalibration of pressure sensor readings based on zero-pressure baseline analysis or atmospheric pressure readings);
6. Shared data, computation, or communication capabilities that augment the device's capability from a remote location and/or enables multiple disparate devices to be linked in novel configurations (e.g., a cloud-based diabetes management system that couples home glucose testing with self-reported food consumption, insulin injections, and clinician coaching);
7. Graphical user interfaces that may display data, or computed meta-data, in novel ways (e.g., a blood pressure monitor that displays computed mean blood pressures in the form of time and frequency deviations from a target blood pressure).

From a system engineering perspective, "complex systems" are also referred to as Systems of Systems (SoS), in which capabilities of the entire system is perceived as being greater than the sum of the individual components. SoS have been studied in great detail because they affect virtually every industry sector. An entire generalized life cycle methodology has been developed and published to support such endeavors, too (U.S.D.O.S., 2008 and Wiley, 2015).

Unintended Consequences: Understanding and Managing Emergent Behaviors in Systems of Systems

Contemporary System of System Engineering (SoSE) principles also usually factor in the wide range of human–system interactions, because humans can add critical pattern recognition capabilities that exceed the ability of the technology by itself (e.g., a pilot will notice—and avoid—a deer wandering on a runway, a critical safety observation that is not easily accomplished with any existing computer).

Because it is common for SoS definitions to include human interaction, another frequently cited novel—and quite challenging—attribute of these systems is "emergent behaviors," which are applications of the system that were not part of the original design specifications or anticipated uses (Osmundson et al., 2008; Edworthy and Hellier, 2006). For aviation, an example of emergent behavior is pilots napping on long treks while the plane is on autopilot (NTSB reports on pilots falling asleep and Both pilots "slept" for an HOUR as packed airliner flew on for 600 miles), leaving the plane on autopilot in dangerous weather conditions (Loss of Control on Approach, Colgan Air, Inc., Operating as Continental Connection Flight 3407, Bombardier DHC 8 400, N200WQ), or completely disabling low-altitude safety systems, with deadly results (Investigation: Air France 296).

Substantially dangerous emergent behaviors with complex clinical systems are now being studied carefully. One example is nurses observed "riding the guardrails" of smart infusion pumps by skipping dose calculations and, instead, increasing pump dose settings until the pump balks, then reducing the rate just below that threshold (Wetterneck et al., 2006).

Another ongoing and extremely dangerous emergent behavior has been nurses turning off or ignoring medical device alarms, which has had fatal consequences (Instrumentation, A. for the A. of M., 2011 and Sounding the alarm).

In another dramatic study, researchers observed multiple unexpected and potentially dangerous ways nurses routinely defeated medication bar code safety systems (Koppel et al., 2008).

Another example is adolescent diabetes patients who schedule their self-administered glucose tests to hide drinking binges (Jaser et al., 2011).

A final example: many physicians and nurses are finding safe use of computerized physician order entry and electronic medical record systems to very difficult, and the embedded flaws have created serious patient hazards (Koppel et al., 2005). The risks are so serious that US National Institute of Standards and Technology developed standardized training and product testing to help manufacturers, researchers, and government regulatory agencies study, analyze, and address the problem (Health IT Usability).

Management Approaches for 21st-Century Complex Clinical Systems

The convergence of clinical and information technologies has been discussed and documented since the beginning of the 21st century (Sloane, 2000). That convergence is accelerating each year as new technologies like smartphones, apps, and wearable sensors have been applied to healthcare applications. Smartphone and tablet apps have become one of the largest segment of the apps industry, and they are used by physicians, nurses, administrators, and patients.

As documented in FDA, FCC, AAMI, and other reports listed above, one of the results of the convergence is that the resulting in-hospital systems are actually composites of many different vendors software and hardware products, often spanning two or more decades of innovation and refinement. In most hospitals, it is often no longer possible to discern where a "medical device" ends and the "purely business" information technology function begins.

Today, sensors and devices measure physiological parameters of all sorts, translate them to digital messages, transit the messages via whatever wireless or wired networks are available, and store and retrieve the data from large storage area networks which may be offsite, or "in the cloud."

Clinical engineers are collaborating with Systems Engineers to develop, adapt, and/or adopt more appropriate tools and methods to manage the current and emerging systems. Ultimately, the clinical engineer's "toolbox" will need to blend several existing discipline, including:

1. "Classic" medical equipment control and IPM processes, as published by AAMI, ECRI, and ASHE;
2. Statistical Process Control based quality management tools so that Reliability Centered Maintenance is precisely and accurately data-driven;
3. Quality Systems Management that continuously reduce errors, using methods similar to the FDA GMP, Joint Commission's Total Quality Management, and ISO 9000;
4. Computer and communication system risk disclosure, analysis, federation, management, using methods like ISO 80001;
5. Information security management methods (Manufacturer Disclosure Statement for Medical Device Security (MDS2));
6. Product recall and update compliance per manufacturer and/or regulatory body notification;
7. National and regional compliance practice requirements in accordance with Federal, State, and Accreditation regulations and guidelines;
8. Entirely fresh thinking and approaches to deal with emergent behaviors of users with these new complex systems to identify and rectify dangerous, unintended, and unacceptable risks;
9. New models of V-Model System Life Cycle planning and management that embody and adapt the Verification and Validation framework for unit, subsystem, and complete system safety and performance for complex medical systems.

Conclusion

The management of complex clinical systems is a discipline under development. Many of the documents cited above may be downloaded for free, or purchased at a nominal charge. They are becoming essential "reading list" materials for clinical engineers seeking to update their skills and knowledge for the 21st century. An anthology of some of these subjects, with several updated resources, was just published by AAMI in June 2015. Although it only addresses a portion of the topics needed to manage complex clinical systems, it may be a useful resource for newcomers to the field (A Practicum for Healthcare Technology Management).

Several additional valuable comprehensive texts on relevant subjects are available on foundational System Engineering skills are available at no charge for PDF download through

US National Academy Press for the National Academies of Medicine and/or Engineering, including:

1. Mental Models in Human-Computer Interaction: Research issues About What the User of Software Knows (1987) (http://www.nap.edu/openbook.php?record_id=790);
2. Human-System Integration in the System Development Process: A new look (2007) (http://www.nap.edu/openbook.php?record_id=11893);
3. Building a Better Delivery System: A new Engineering/Health Care Partnership (2005) (http://www.nap.edu/openbook.php?record_id=11378).

Clinical engineers in the 21st century are faced with a significant opportunity and choice: either expand the discipline's skills, capabilities, and scope of work, allow future clinical technologies to be managed by a new discipline of Health Technology Management (HTM Resources), or surrender authority and control to other C-Suite leadership, such as the CIO, CMIO, or CTO.

The first option, expanding the Clinical Engineering discipline, is the most likely to provide comprehensive solutions for healthcare, but it certainly will not be the path of least resistance.

References

Edworthy, J., Hellier, E., 2006. Alarms and human behaviour: implications for medical alarms. Br. J. Anaesth. 97, 12–17.

Guide, U.S.D.O.D.S., 2008. Systems Engineering Guide for Systems of Systems. Version 1, 20301–23090.

Instrumentation, A. for the A. of M., 2011. Clinical alarms: 2011 summit. Association for the Advancement of Medical Instrumentation.

Jaser, S.S., Yates, H., Dumser, S., Whittemore, R., 2011. Risky business risk behaviors in adolescents with type 1 diabetes. Diabetes Educ. 37, 756–764.

Koppel, R., Metlay, J.P., Cohen, A., Abaluck, B., Localio, A.R., Kimmel, S.E., et al., 2005. Role of computerized physician order entry systems in facilitating medication errors. JAMA 293, 1197–1203.

Koppel, R., Wetterneck, T., Telles, J.L., Karsh, B.-T., 2008. Workarounds to barcode medication administration systems: their occurrences, causes, and threats to patient safety. J. Am. Med. Inform. Assoc. 15, 408–423.

Olansen, J.B., Rosow, E., 2001. Virtual Bio-instrumentation: Biomedical, Clinical, and Healthcare Applications in LabView. Pearson Education.

Osmundson, J.S., Huynh, T.V, Langford, G.O., 2008. KR14 Emergent Behavior in Systems of Systems. In: INCOSE International Symposium. Wiley Online Library, pp. 1557–1568.

Ridgway, M., Atles, L.R., Subhan, A., 2009. Reducing equipment downtime: a new line of attack. J. Clin. Eng. 34, 200–204.

Rosow, E., Adam, J., Coulombe, K., Race, K., Anderson, R., 2003. Virtual instrumentation and real-time executive dashboards: solutions for health care systems. Nurs. Adm. Q. 27, 58–76.

Sloane, E.B., 2000. The golden age of clinical engineering: digital convergence and IT partnerships. Biomed. Instrum. Technol. 35, 95–103.

Wang, B., Sloane, E.B., Patel, B., 2001. Quality management for a nationwide fleet of rental biomedical equipment. J. Clin. Eng. 26, 253–269.

Wetterneck, T.B., Skibinski, K.A., Roberts, T.L., Kleppin, S.M., Schroeder, M.E., Enloe, M., et al., 2006. Using failure mode and effects analysis to plan implementation of smart iv pump technology. Am. J. HealthSyst. Pharm. 63, 1528–1538.

Wiley, 2015. INCOSE Systems Engineering Handbook: A Guide for System Life Cycle Processes and Activities. John Wiley & Sons.

Web References

A Practicum for Healthcare Technology Management. Available from: http://www.aami.org/productspublications/ProductDetail.aspx?ItemNumber=2431.

Boeing Improves 777 Maintenance Program. Available from: http://boeing.mediaroom.com/2006-05-04-Boeing-Improves-777-Maintenance-Program.

Both pilots 'slept' for an HOUR as packed airliner flew on for 600 miles. Available from: http://www.dailymail.co.uk/news/article-1222401/Pilots-suspended-falling-asleep-overshooting-airport-150-miles.html.

British Airways Las Vegas plane fire: Eyewitnesses describe blaze. Available from: http://www.bbc.com/news/world-us-canada-34195463.

Defibrillator: AO Pulsar 4. Available from: http://www.emsmuseum.org/virtual-museum/by_era/articles/399002-Defibrillator-AO-Pulsar-4.

ETOPS-Based Maintenance Guidance. Available from: http://www.boeing.com/commercial/aeromagazine/aero_05/textonly/m02txt.html#etops.

FAQS for the wireless challenge in healthcare. Available from: http://s3.amazonaws.com/rdcms-aami/files/production/public/FileDownloads/HT_Wireless/140829_Wireless_FAQs.pdf.

FCC/FDA Joint Meeting on Life Saving Wireless Medical Technology Day-1. Available from: https://www.fcc.gov/events/fccfda-joint-meeting-life-saving-wireless-medical-technology-day-1.

Health IT Usability. Available from: http://www.nist.gov/healthcare/usability.

Healthcare technology in a wireless world. Available from: http://s3.amazonaws.com/rdcms-aami/files/production/public/FileDownloads/Summits/2012_Wireless_Workshop_publication.pdf.

HTM Resources. Available from: http://www.aami.org/membershipcommunity/content.aspx?ItemNumber=1485http://www.theamericannurse.org/index.php/2013/11/04/sounding-the-alarm.

IHE Patient Care Devices. Available from: www.IHE.net/Patient_Care_Devices.

Interoperability Specification. Available from: http://hitsp.org/ConstructSet_Details.aspx?&PrefixAlpha=1&PrefixNumeric=77.

Investigation : Air France 296. Available from: http://www.airdisaster.com/investigations/af296/af296.shtml.

Loss of Control on Approach, Colgan Air, Inc., Operating as Continental Connection Flight 3407, Bombardier DHC 8 400, N200WQ. Available from: http://www.ntsb.gov/investigations/AccidentReports/Pages/AAR1001.aspx.

Manufacturer Disclosure Statement for Medical Device Security (MDS). Available from: http://www.himss.org/resourcelibrary/MDS2.

Medical Device Data Systems, Medical Image Storage Devices, and Medical Image Communications Devices. Available from: http://www.fda.gov/downloads/MedicalDevices/DeviceRegulationandGuidance/GuidanceDocuments/UCM401996.pdf.

Medical drevice interoperability. Available from: http://s3.amazonaws.com/rdcms-aami/files/production/public/FileDownloads/Summits/2012_Interoperability_Summit_Report.pdf.

Multi-Engine Maintenance. Available from: http://www.boeing.com/commercial/aeromagazine/aero_05/textonly/m02txt.html.

NTSB faults Asiana pilots for 777 crash, says complex automated controls a factor. Available from: http://www.seattletimes.com/business/ntsb-faults-asiana-pilots-for-777-crash-says-complex-automated-controls-a-factor/.

NTSB reports on pilots falling asleep. Available from: http://abcnews.go.com/Travel/story?id=5042619.

Sounding the alarm. Available from: http://www.theamericannurse.org/index.php/2013/11/04/sounding-the-alarm.

Technical note. Available from: http://hitsp.org/ConstructSet_Details.aspx?&PrefixAlpha=5&PrefixNumeric=905.

The Acquisition and Maintenance of Medical Equipment. Available from: https://ia801605.us.archive.org/1/items/acquisitionmaint00unit/acquisitionmaint00unit.pdf.

Unique Device Identification—UDI. Available from: http://www.fda.gov/MedicalDevices/DeviceRegulationandGuidance/UniqueDeviceIdentification/.

Wireless technology. Available from: http://www.aami.org/newsviews/content.aspx?ItemNumber=1395.

Decision Support Systems in Healthcare

Ernesto Iadanza[1], Gabriele Guidi[2], Alessio Luschi[2]

[1]CHAIRMAN, CLINICAL ENGINEERING DIVISION OF THE INTERNATIONAL FEDERATION FOR MEDICAL AND BIOLOGICAL ENGINEERING (IFMBE); EDUCATION AND ACCREDITATION COMMITTEE OF THE INTERNATIONAL UNION FOR PHYSICAL AND ENGINEERING SCIENCES IN MEDICINE (IUPESM), DEPARTMENT OF INFORMATION ENGINEERING – UNIVERSITÀ DI FIRENZE, ITALY, FIRENZE, ITALY [2]DEPARTMENT OF INFORMATION ENGINEERING, BIOMEDICAL LAB, FIRENZE, ITALY

Introduction

Decision Support System (DSS) is the common name used for referring to tools that can help in supporting decision-making and problem solving. These technologies date back to the 1960s, in many fields of science. Today DSSs are computer technology solutions widely applied also to healthcare, both in hospital management/administration and in clinical practice. Classic DSS tools include:

- database management capabilities for accessing data and knowledge (internal or external)
- an engine that elaborates this data for helping the decision makers, often including state-of-the-art machine learning techniques
- an user interface designed for enabling interactive queries, reporting and graphing functions (Shim et al., 2002).

To date, these systems increasingly are based on distributed models that include connecting capabilities for taking advantage of the power of the Internet in all their functions. A modern DSS is enabled to take advantage of the Web connection in all its functionalities. It can access data from devices and databases from all over the world (Open Data), make the most of tremendous computing capacities based on grids and other distributed computation techniques, as well as including web-based interfaces for collaborative use even on mobile devices.

A good view on how the Web affects the DSS design, functions, models, and architectures is proposed by Hemant K. Bhargava et al. in "Progress in Web-based decision support technologies" (Bhargava et al., 2007).

Clinical Engineering. DOI: http://dx.doi.org/10.1016/B978-0-12-803767-6.00007-6

Healthcare is increasingly including the use of DSSs in all its processes, both clinical and nonclinical. This also affects the governance structure and the organization design. According to Rajalakshmi: "Decision making in healthcare is primarily done in two areas. The first area (lower level) involves patient management, diagnosis and treatment, record keeping, finance and inventory management. The other area involves higher level decision making that gives the hospital a competitive edge. The shareholders that play an important role in lower level decision making constitute of doctors and nurses" (Rajalakshmi et al., 2011).

An interesting analysis of the state-of-the-art in Clinical Decision Support Systems (CDSSs) has been prepared in 2009 for the Agency for Healthcare Research and Quality. The report describes the various types of CDSSs and their impact and effectiveness on the clinical practice, as well as their design and implementation (Berner, 2009).

Bright et al. published a recent systematic review on the effect of CDSSs in the *Annals of Internet Medicine* in 2012 which the reader can eventually refer to. The review evaluates the effect of CDSSs on clinical outcomes, healthcare processes, workload and efficiency, patient satisfaction, and costs (Bright et al., 2012).

When it comes to clinical engineers (CE), most of the problems can be seen as multi-attribute/criteria decision problems, which means evaluating a finite number of alternatives and criteria for taking a decision. DSS can also be a valid support for these processes, giving the CE a valid tool both for taking decisions and for involving the top management in the decision process. The complexity of the available systems, some commercially available and some custom designed, can vary a lot. An analysis of the literature, proposed by Kamel et al. in "Decision Support Systems in Clinical Engineering," shows the adoption of the following methods (Kamel and Tawfik, 2010):

- Pros and Cons Analysis
- Kepner-Tregoe (K-T) Decision Analysis
- Cost-benefit Analysis
- Multi-Attribute Utility Theory (MAUT)
- Analytic Hierarchy Process (AHP)

A Clinical Decision Support System for Congestive Heart Failure Management

In this section, we will describe an example of CDSS that we designed and implemented for facilitating the decision process in the management of the Congestive Heart Failure.

Chronic heart failure (CHF or HF) is a cardiac condition characterized by heart inability to supply blood in adequate quantities to meet the organism needs. HF is the result of other diseases, such as heart valves diseases, cardiomyopathies, congenital heart disease, arterial hypertension, abuse of alcohol and drugs, coronary artery disease, and heart attack. Considering that HF is prevalent in older people with much comorbidity, and that the diseases that cause it are often multiple, it is a strongly multifactorial disease. HF is a common disease and it is very expensive for the health system: it affects 3−20 out of

1000 adults and up to 10% of people aged between 80 and 89 (Takeda et al., 2012). Once HF has been diagnosed, drug therapy is the mainstay of treatment. However, a structured management and monitoring of HF patients is also essential for improving the consequences of the disease. Until a few years ago, chronic patients were managed with the so-called waiting medicine rule, meaning that the treatment was only performed once the patient showed symptoms of exacerbation. More recently, with the advent of the Chronic Care Model (CCM) (Wagner et al., 2001), the health system is increasingly adopting the "initiative medicine" code, according to which, patients should be monitored and treated before the disease worsen to its acute phase. Building an effective disease management strategy requires the analysis of many variables, including the care setting, patient and family ability to perform self-management, and the severity of the disease (Inglis et al., 2011). Some Cochrane Collaboration reviews show the effectiveness of preventing HF-related re-hospitalization through integrated assistance schemes and parametric monitoring, rather than just drug therapy (Takeda et al., 2012; Inglis et al., 2011). In addition, current practice guidelines and recent consensus statements on HF, agree that HF is a multifactorial disease that requires new models of coordination between health professions and patients (proactive self-management) in order to improve care continuity outside the hospital (Yancy et al., 2013). In this context, with a highly multifactorial disease requiring a strong management complexity, a CDSS may be particularly useful. This chapter describes the system *Heart M&M* (Heart Failure Management and Monitoring) that proposes a decision-supported cooperative framework for HF monitoring involving multiple stakeholders by combining manageability features with artificial intelligence algorithms.

The Cooperative Framework

This subsection explains the ideal cooperative framework in which the system *Heart M&M* is intended to operate, because a DSS is most effective when working in the conditions of "intended use" it was designed for: in this case, this means HF patients monitoring scenario structured in three levels. Each level of monitoring is characterized by two indices: the possibility of frequently measuring parameters (*Freq*) and the possibility of measuring parameters strongly correlated with HF (*Rel*). These two conditions can hardly coexist in real scenarios and therefore an exhaustive monitoring is composed by the combination of all the three levels, which are:

- Level 1: *Scheduled follow-up* in clinic (or point of care)—The patient goes by appointment to the clinic where the cardiologist makes an exhaustive visit acquiring many parameters/information such as: Height and weight (Body Mass Index, BMI), Systolic and Diastolic Blood Pressure, Heart Rate, Oxygen Saturation, Ejection Fraction (EF), Brain Natriuretic Peptide (BNP), Bioelectrical Impedance Vector (BIVA) Parameters, NYHA Class, 12-lead EKG report (e.g., presence of bundle branch block, tachycardia, atrial fibrillation),

Etiology, Comorbidity, Current Pharmaceutical and Surgical Therapy (pacemaker or ICDICD/CRT). This type of monitoring is characterized by:

- *Freq*: *low*, one visit every 6 months
- *Rel*: *high*, since it is a complete visit made by the cardiologist, some strong HF-related parameters and information can be acquired as the BNP, the EF or the situation of comorbidity.

- Level 2: *Home monitoring* performed by nurse—A nurse periodically goes to the patient's home with a measurement kit. He performs a physical examination and acquires some parameters, that are: Weight, Blood Pressure, Oxygen Saturation, Bio-Impedance using portable instrument, Capillary Dosage of BNP using portable device. This type of monitoring is characterized by:
  - *Freq*: *medium*, the nurse goes to the patient every 15 days.
  - *Rel*: *medium*, the visit is not exhaustive because not made by a specialist but still includes useful and strong correlated parameters such as BNP capillary and bioelectrical impedance.

- Level 3: *Patient self-monitoring*—This scenario is intended to set up an easy-to-use measuring kit to allow self-measurement on a daily basis at patient's home. Parameters that the patient will be able to self-measure are 2-lead Electrocardiography (EKG), Bio-Impedance signals, the measurement of the Pulse Transit Time (PTT—the time interval between the cardiac contraction, EKG's R-wave, and the arrival of the blood wave to the periphery), Blood Pressure, Oxygen Saturation, and Weight. This type of monitoring is characterized by:
  - *Freq*: *high*, the patient may self-measure parameters daily.
  - *Rel*: *low*, the parameters easily measurable by the patient himself are not strong HF markers.

The *Heart M&M* System

The *Heart M&M* system is designed and trained to assist the decisions of clinician stakeholders involved in this HF monitoring setting described in previous chapter (Figure 7.1) who are cardiologists (level 1) and nurses (level 2). For the system to be useful, the two different stakeholders need distinct tools and a decision support functionality, because suggestions that can be obvious to a cardiologist can be very useful for a nurse when he is at patient's home (monitoring level 2). The whole *Heart M&M* system consists of an ICT infrastructure that coordinates: a desktop application for the cardiologist, a mobile application for the nurse, and a core of Machine Learning Algorithm that collects data and provides different outputs. The DSS function is made possible by the intelligence resulting from the core of machine learning and the way it is inserted in the monitoring workflow, facilitated by the two management tools. In this context, the HF-DSS should be considered as the triplet of special purpose management tools, intelligent advice by machine learning service, and operational workflow. The data collected from patient self-monitoring are also connected to the system. These data are acquired with devices and sent by remote transmission to be monitored and shown to the doctor. This last part is still work in progress in the *Heart M&M* system and currently the DSS does not include the information acquired from level 3 to provide aid in clinical decisions.

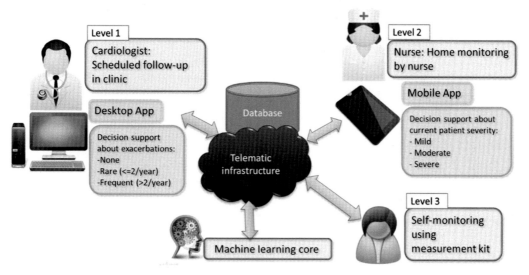

FIGURE 7.1 *Heart M&M* system schema.

The Machine Learning Core

The machine learning core consists of a Random Forest (RF) (Breiman, 2001; Guidi et al., 2013), properly trained to provide information to:

- the cardiologist, in the outpatient setting, helping to manage the patient;
- the nurse, who performs the home visit, helping to assess the severity of the current state of the patient.

The RF was then trained to provide two types of input: a three level prediction of exacerbations that the patient may have during a year (*None, Rare < = 2* per year, or *Frequent > 2* per year), and a three level assessment of HF severity (*mild, moderate, severe*). Both outputs are obtained by training the RF in a supervised mode (Figure 7.2), by providing as input some example of the phenomenon, that is a pattern of parameters (patient parameters), together with the desired output (i.e., the training target, respectively, *none, rare, frequent* and *mild, moderate, severe*).

In the training phase, the desired outputs are entered by the cardiologist through a special interface and, next, cardiologist and nurse will use the trained system, providing new parameters and obtaining the corresponding desired outputs, useful to assist their decisions.

In the HF-severity output training, which will be helpful to the nurse during home visits, the cardiologist in his outpatient visits acquires the parameters of each patient and assigns to this parametric pattern the value of *mild, moderate,* or *severe* depending on the current HF state. This process allows the system to learn managing the severity assessment in a multiparametric mode rather than basing on some thresholds for each parameter.

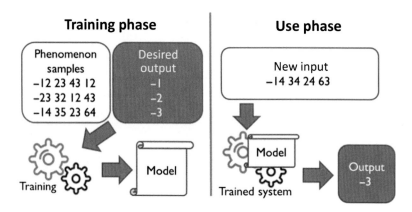

FIGURE 7.2 Supervised training schema.

Table 7.1   Dataset Distribution

Type of Output	No. of Patients in Class 1	No. of Patients in Class 2	No. of Patients in Class 3	Sum
CHF Severity (mild/moderate/severe)	93	92	65	250
CHF decompensation (none/rare/frequent)	161	55	64	250

Table 7.2   Results of DSS Providing Output About Complexity of Care

	Accuracy	No. of Critical Errors 1-3	No. of Critical Errors 3-1	"None" vs All		"Rare" vs All		"Frequent" vs All	
				Sens	Spec	Sens	Spec	Sens	Spec
Average in 10-folds	71.9%	3 (sum)	2 (sum)	0.57	0.79	0.65	0.60	0.59	0.96

For the output indicator of possible exacerbations during the year, which is helpful to the cardiologist himself, the training is made on a retrospective basis using patient historical data. At the beginning of the year, the doctor assigns to each parametric-pattern, related to the previous year, the label *"none, rare, or frequent"* indicating the number of exacerbations. In this way, the output gains predictive *value* and, once trained, the RF provides the cardiologist information about the future complexity of a patient care during the outpatient visit. The training dataset is shown in Table 7.1, while the results of the training process in 10-fold cross-validation are shown in Tables 7.2 and 7.3 where accuracy, in a three-class classificator, means:

$$Accuracy = \frac{\sum_{i=1}^{N^\circ Class} \frac{TP_i + TN_i}{TP_i + TN_i + FP_i + FN_i}}{N}$$

Table 7.3   Results About DSS Providing Output About Severity Assessment

	Accuracy	No. of Critical Errors 1-3	No. of Critical Errors 3-1	"Mild" vs All		"Moderate" vs All		"Severe" vs All	
				Sens	Spec	Sens	Spec	Sens	Spec
Average in 10-folds	81.3%	0	1 (sum)	0.75	0.84	0.67	0.80	0.87	0.95

and sensitivity and specificity are calculated with the "one vs all" method (TP = True Positive, FP = False Positive, TN = True Negative, FN = False Negative).

Critical errors in Tables 7.2 and 7.3 are defined as cases in which a patient who had no decompensation (class *none*) or was in severity class *mild*, was wrongly classified as one with *frequent* decompensation or as a *severe* CHF (critical error 1-3). Vice versa, critical errors 3-1 were defined as cases in which *frequent* decompensation or *severe* CHF was erroneously classified as *none* or *mild*, respectively.

Considering that it is a classifier on three classes, these are good results, especially for the low number of critical errors of both type 1-3 and type 3-1 and the high specificity obtained for both the most severe classes (Frequent and Severe), avoiding to generate false alarms.

Cardiologist Front-End

As described above, the machine learning core is only a part of the whole process of clinical decision support, in which also management tools and various HF-specific functionality contained in them have a very important role.

The cardiologist front end, that is a desktop app, has the threefold function to serve as:

- HF special purpose management tool, integrating some disease-specific features, and allowing doctor to check information of the monitored patients
- Interface to provide useful information, for the supervised training
- Interface to receive decision support by the trained machine learning core.

In Figures 7.3, 7.4, 7.5, and 7.6 are shown some screenshots of the cardiologist front-end including simulated patients and data. Since the software is experimentally used by Italian cardiologists, text in screenshots is mostly in Italian language.

In particular, Figure 7.3 shows the software start screen, where the cardiologist can select the patient from a list (also by text search) if already existing in the database, or add a new one.

Once selected a patient, Figure 7.4 shows the input form of the parameters acquired during the home visit. Note how, in this screen, the cardiologist may also associate with the parametric pattern inserted the training target for the severity assessment output (mild−moderate−severe) that will be useful to the nurse.

In a specific section of the cardiologist front-end, we integrated implementations of four literature HF prognostic models, including the well-known Seattle Heart Failure Model

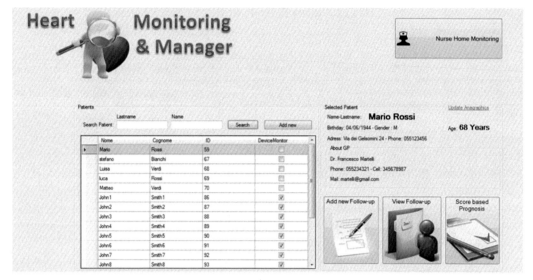

FIGURE 7.3 Cardiologist front-end—patient selection.

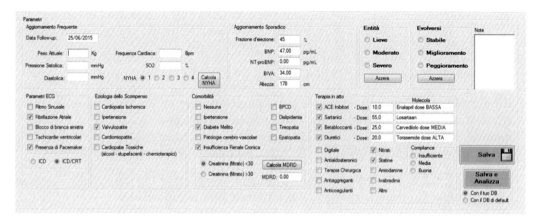

FIGURE 7.4 Cardiologist front-end—parameters input form.

(SHFM) [7]. Figure 7.5 shows the interface to automatically calculate these prognostic score for each patient using the information contained in the database.

Finally, Figure 7.6 shows the control panel associated with each patient, through which doctor can graphically display the trend of the parameters in the various follow-up visits as well as select from the list on the left a specific follow-up and display details about it. Also through this interface (in bottom left), cardiologist can benefit from the advice provided by the machine learning core about the complexity of patient care.

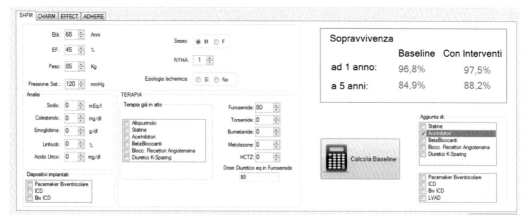

FIGURE 7.5 Cardiologist front-end—literature score based models form.

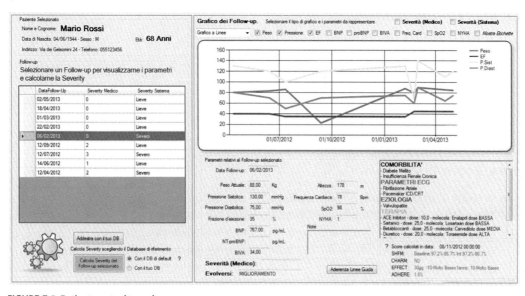

FIGURE 7.6 Patient control panel.

Many other expedients and management tricks that aim to simplify the clinical practice and the parameters input have been implemented, such as:

- Real-time checks and alerts on each input field (parameter out of range, empty field, wrong format)
- In the therapy section, a feature establishing the appropriate molecule and if the prescribed dose is low, medium, or high for that molecule simply basing of the inserting dosage of certain categories of drugs (ACE inhibitors, beta-blockers, diuretics, and ARBs)

FIGURE 7.7 Mobile app—selection of existing follow-up.

- An automatic calculation of some indices, such as body mass index, MDRD in case of kidney failure
- A graphical view of drug dosage that helps the doctor to decide for "a gradual increment in dose if lower doses have been well tolerated" for certain categories of drugs, as recommended by the guidelines (Yancy et al., 2013).

Nurse Front-End

Font-end for nurses is used during home visits in the level 2 of monitoring in order to acquire patient parameters and consequently obtain the response of the machine learning core about the current state of the patient severity (Mild, Moderate, or Severe). To use it at patient's home, the front end is therefore a mobile application that can be used both on smartphones and on tablets. Figures 7.7 and 7.8 show two screenshots of the application containing a simulation that runs on a 10-inch tablet, respectively, for the selection of an existing follow-up related to a patient, and for the addition of a new follow-up.

A DSS for Hospital Administration and Facilities Management

In this section, we will show an example of a nonclinical DSS that we realized for hospital top managers. The main module of the Computer-Aided Facilities Management (CAFM) system is designed to manage and analyze digital plans of hospital buildings, coded on specific layers. The system maps hospital's inner organization, destinations of use, and

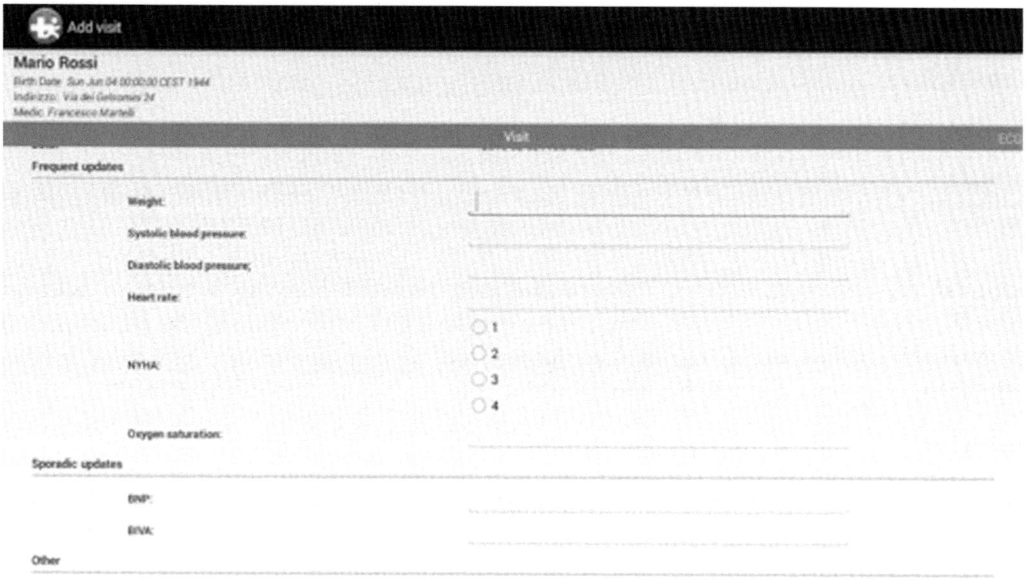

FIGURE 7.8 Mobile app—add a new follow-up.

environmental comforts giving quantitative, qualitative, and graphical reports. The core database is linked to other existing databases, in order to use the system as a central control cockpit. Outputs can be used by top management as a decision-support tool in order to improve hospital's structure and organization and to reduce the major workflow risks. Furthermore the suite has many plug-ins that contribute to a complete analysis and management of the healthcare facilities, which will be highlighted later through the chapter.

Modern hospital organization is related to a big amount of data, about structures, facilities and workers, that need to be organized and set in functional relations among them. Today hospitals must undergo a big quantity of requirements in order to fulfill their clinical and medical duties. These requirements are set by national and international institutions, which force structures to respect given parameters in order to have good hygienic, qualitative and organizational standards granted.

Clinical Engineering Services and Technical Departments must deal with these issues and find solutions to fully satisfy technological, structural and organizational needs for such a complex structure as a hospital, using different technical tools to monitor the hospital by measuring quantitative, architectonical, technological and people-related parameters.

Many of these systems are based on special hospital dedicated DataBase Management Systems (DBMS) and Building Information Modeling tools (BIM). Data can be stored and then aggregated in different ways to answer to a wide range of queries.

Computer-Aided Facility Management (CAFM) systems are decision-support tools based upon Integrated Healthcare Facility Management Models (IHFMMs) which provide indexes on those processes that can affect the performance of the healthcare structure. These tools can be very useful for the top management in performance and risk evaluations, business management, and development (Rodriguez et al., 2003; Iadanza et al., 2007; Lavy and Shohet, 2007).

Workplace Management Systems (WMSs) are solutions designed to manage real estate facilities, allowing users to assess, analyze, and reorganize the company assets in order to preserve their value, improve their effectiveness, and respond to multiple needs. They provide access to stored information regardless of the workplace: data and plans can be acquired through web services, using a common browser over an Internet or intranet network. These systems drive a Computer-Aided Design (CAD) engine to store information about space-units, assets, biomedical equipment, plants, phone and data plugs and wiring, employments, providing visual outputs. Data are stored in a dedicated database; map files are just linked. The main informative unit, i.e., the maximum degree of detail, may be the Homogeneous Functional Area (a set of rooms pooled together by destination of use) or the room itself: the first approach offers an useful overall view but does not allow accurate information on single rooms supplies; the second one is considered to be a more full-scale method.

The proposed system is a suite that consists of a main software module and several extra tools. They all refer to the same inner database and have links to the Hospital Information System (HIS). A stand-alone main executable application (SACS) monitors the status quo of the buildings referring to beds, square meters, destination of use, functional areas, and many other features for every room by driving DWG maps in Autocad. A rendering engine is dedicated to creating fully HTML5-compliant SVG files to show the information allowing several ways for data aggregation (Iadanza, 2009; Luschi et al., 2014; Luschi et al., 2015).

The system is also provided with a web-based search engine, named EUREKA, developed in ASP.NET that allows users to perform free-text queries providing real-time reports. One more module, an ASP.NET web application called LICENSE, is an expert system capable of automatic verification of the compliance with structural and technological legislative requirements (Guidi et al., 2015).

SACS Main Module

The core-engine software drives CAD maps to realize an "everything inside DWG" system: all data are stored inside the maps through the Extended Data (XData) properties of the AutoCAD Object. The drawings are self-sufficient and the information can be always rebuilt using nothing but the CAD files. A SQL Server 2008 database that contains support tables is still used to store date in order to perform queries and aggregations on them. The XData propriety allows heterogeneous type of data to be stored inside the DWG file, directly linked to CAD objects (polyline, hatches, lines, blocks, etc.): it is possible to store strings, numbers (both floating and integer), times/dates, and binary data.

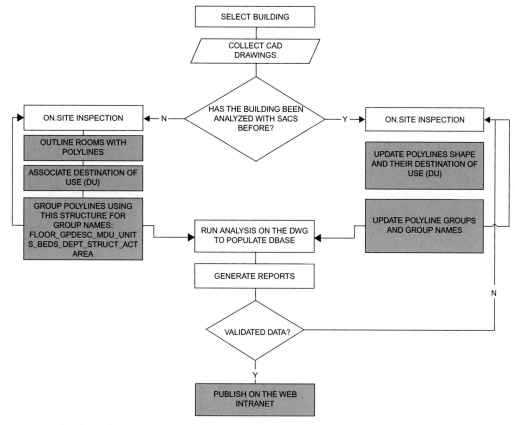

FIGURE 7.9 Flowchart of SACS operative steps.

The current release of the software analyzes the plan of the hospital and gives information about dimensions and aggregation of rooms and spaces, structured in destination of use (DU) and main destination of use (MDU). Data are collected through on-site surveys and personnel interviews. Spaces are classified by their usage and by customer expectations in terms of environmental comfort. Survey information is then data-entered into SACS and linked to the DWG AcadPolyline objects that outline the perimeter of a room. Polyline color has been used as a parameter to identify DU and MDU and it is associated with the intensity of care for each DU. The software has a module to make multiple selections on the DWG file and associate to polylines an RGB color for each of the 42 available destinations of use coming from the database (Figure 7.9).

The difference between DU and MDU is tiny but essential. The Destination of Use of a room is the effective description of the activity carried out (Intensive Care Unit, Ward, Ambulatory, Day Hospital, Day Surgery, etc.) while the Main Destination of Use is the description of the functional connection among a group of rooms, even with different DUs (an ensemble of rooms classified as MDU = 'Ward' will join, together with the actual ward room, other different typologies such as hallways, inner and outer washrooms, warehouses etc.). This parameter allows considering each room in its real operative context.

The 42 DUs have been sub-categorized in 246 sub-destinations or classes to allow a more detailed complexity of the surveys. For example, the Operating Room DU assembles heterogeneous areas, each having its articular requirements (patient filter zone, operators filter zone, etc.): the classes allow the user to specify these particular parameters for each room.

The following listed data is also managed by SACS in order to give the user more information about the analyzed spaces:

- ROOM CODE, following the scheme BLD_LEVEL_ROOM, where BLD is the code of the building, LEVEL is the number of the floor mapped by the linked DWG file (-1 for the basement, 0 for the ground floor, 1 for the 1st floor, etc.), and ROOM is a three-digit formatted increasing number. The joining of these three codes gives an unique explicit code that identifies every single room inside the hospital showing the right position;
- ATU: a code of the Air Treatment Unit (ATU) that feeds the room;
- EG: Electrical Group for the medical rooms, that can be 0, 1, or 2 (X-unused for the non-medical rooms);
- NGAS: the number of medical gasses connection detailed for air, oxygen, nitrous oxide, carbon dioxide, and vacuum terminals;
- H: the room height;
- SLR: the square-light ratio.

The whole information in each DWG file is always extracted by the software itself and inserted in an SQL Server 2008 support table linked via ActiveX Data Objects (ADO) in order to assure redundancy backup functionalities.

SVG Rendering Engine

This is an ASP.NET web application that allows users to dynamically access information and plants about the hospital buildings with no need to have a DWG rendering engine installed on their own computers. The application allows user to visualize the last updated map of the requested floor of a building in real time by using AJAX and JQuery. Every time a click event on a room is performed, it will trigger an asynchronous access to the server database which will load the updated information without any callback to the page itself, to spare computing time. The engine provides the basic functions of a CAD engine such as multi-selection, panning, zooming, text placement and scaling, multiple selection by DU, Operative Units, or Departments. Besides it highlights a room when the user clicks on the icon representing a person that is located in that room.

Some kind of information (like health technologies and costs) are protected with a password, so that only authorized users can access them.

Figures 7.10 and 7.11 show the described functionalities.

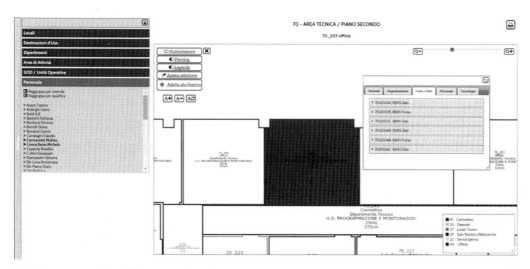

FIGURE 7.10 Screenshot of the SVG Rendering Engine. Pan, zoom, and multi-selection functions can be shown in the top-left corner of the map. Destination of Use, Department, Operative Unit, and Personnel can be shown in the top-left frame of the window. General, Organizational, Plants and Technological Information are on the bottom-left frame.

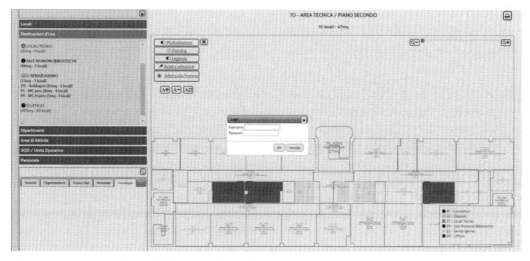

FIGURE 7.11 Access to protected information (technological ones in the example) are protected with a password.

LICENSE

One more key module of the DSS is LICENSE. It is an expert system that supplies a decisional support for assessing the hospital compliance to the desired requirements. It is used to verify the conformity of the structure to the national and regional legislative standards and

Table 7.4   LICENSE Standardization Requirement Stored in the Database

ID	COND_PL	DU	ESPRESSIONE	GRUPPO
8	>1	24	9 + 7*([TB_QUERY.PPLL]-1)	1

to manage deficiency, to schedule the interventions by their priorities and to perform a risk analysis in order to reduce the major risks in accordance with the available resources (Iadanza et al., 2014).

The tool uses the data stored in SACS to evaluate the structural and plant requirements, while interfacing with the hospital health technologies database. As structural standards the requested parameters are the DUs of the rooms, their surfaces, the number of beds, and the presence of certain typologies of rooms (identified through their Class) that must exist at the floor and that must be assigned to the same activity area which is being analyzed.

In terms of technological requirements, the assessment of the compliance is based on the presence of a given technological asset. This checking is made possible by searching for a device through the first three digits of its CIVAB class among a pre-filtered set according to the analyzed activity area. The CIVAB encoding is the prevalent used categorization in Italy, which allows unique identification of biomedical technologies by using an eight alphanumeric characters code that identifies:

- the class (for example, TAC is for computed tomography):
- the manufacturer;
- the specific model.

In order to make the tool as flexible and adaptable as possible, following any legislative changes, every expression and formulas used for the evaluation directly is stored by LICENSE inside the database and can be easily updated (see the example in Table 7.4).

Once the user selects the building, the floor and the activity that need to be evaluated, the system will lead through the filling of the related forms. Each question is automatically answered by the software itself (if possible) evaluating the available data about technologies, plants and structures through the formulas stored in the database. Hence the user can confirm or deny the system response and, in the event it is negative, he can plan an intervention to achieve a requirement fulfilling in accordance with the law (Figure 7.12).

EUREKA

EUREKA is a vital part of the whole DSS. It allows free-text queries on the database. The search is made throughout all the fields related to the acquired data and can be even narrowed by using an advanced search tool to restrict the output (Figure 7.13).

The search algorithm allows users to perform queries using the common Google-like syntax and ASCII characters. Plus it uses a Fuzzy Dictionary to offer the "Did you mean" functionality. EUREKA also provides a dynamic reporting function, which allows users to output

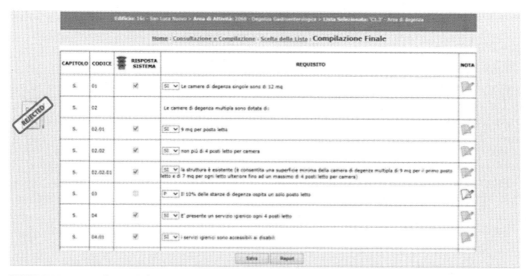

FIGURE 7.12 LICENSE last web-form. The third field represents the system response, while the fourth field is the user response. In case of a P answer by the user, the Note field can be filled to write down an intervention plan. The Rejected icon on the left represents the global state of the form (it results approved once all the requirements are positively answered).

FIGURE 7.13 Example of a query performed via Eureka. In the example, "hematology" has been used as a free-text search query.

reports basing on the search results, with different levels of aggregation and arranged by different scale factors, such as building, building/floor, department, destination of use.

The tool is used every day by different hospital offices, especially by the Top Management, the Clinical Engineering Service and the Technical Department; dedicated personnel frequently update the tool.

The software allows queries on rooms and gives numerical and graphical reports output, therefore becoming a support tool for the healthcare planning. A list of few examples, grouped by users:

- Directors or Nurse Coordinators use the system to know the "spreading" of their units/departments and which ones they possibly have to coexist with.
- Firemen query the system to know the escaping pathways along the buildings, where the fire-escapes and fire-stairs are and which places are more sensible and thus ask for more attention (rooms with combustive agents like ward or Intensive Care Units, super-magnet room of a Magnetic Resonance Imaging, etc.).
- Technical staff uses it almost every day to retrieve parameters used for managing the hospital and to calculate indicators for quality of service or accreditation requirements (by using LICENSE).
- Everybody inside the hospital can query EUREKA to know building's code and name or any medical or nonmedical activities done inside of a given room by knowing only its number. This function is more expanded by introducing information about the personnel so that it becomes an useful tool for people-finding too.

The system is used as a Decision Support System for many purposes and in many different scenarios like transfer management, accreditation requirements assessment, electromedical devices management, general designing and remodeling (Miniati et al., 2010).

References

Berner, E., 2009. Clinical Decision Support Systems: State of the Art. AHRQ Publication.

Bhargava, H.K., Power, D.J., Sun, D., 2007. Progress in Web-based decision support technologies. Decis. Support Syst. 43, 1083–1095.

Breiman, L., 2001. Random forests. Mach. Learn. 45, 5–32.

Bright, T.J., et al., 2012. Effect of clinical decision-support systems—a systematic review. Ann. Intern. Med. 157, 29–43.

Guidi, G., Pettenati, M.C., Miniati, R., Iadanza, E., 2013. Random forest for automatic assessment of heart failure severity in a telemonitoring scenario. Engineering in Medicine and Biology Society (EMBC), 2013 35th Annual International Conference of the IEEE. IEEE, pp. 3230–3233.

Guidi, G., Luschi, A., Miniati, R., Iadanza, E., 2015. EUREKA: a web based search engine for hospitals. 6th European Conference of the International Federation for Medical and Biological Engineering. Springer International Publishing, pp. 625–628.

Iadanza, E., 2009. An unconventional approach to healthcare (Geographic) information systems using a custom VB interface to Autocad, In: BIOSTEC Workshop MobiHealthInf. pp. 13–19.

Iadanza, E., Dori, F., Gentili, G.B., Calani, G., Marini, E., Sladoievich, E., et al., 2007. A hospital structural and technological performance indicators set. 11th Mediterranean Conference on Medical and Biomedical Engineering and Computing 2007. Springer, pp. 752−755.

Iadanza, E., Ottaviani, L., Guidi, G., Luschi, A., Terzaghi, F., 2014. LICENSE: web application for monitoring and controlling hospitals' status with respect to legislative standards. XIII Mediterranean Conference on Medical and Biological Engineering and Computing 2013. Springer, pp. 1887−1890.

Inglis, S.C., Clark, R.A., McAlister, F.A., Stewart, S., Cleland, J.G.F., 2011. Which components of heart failure programmes are effective? A systematic review and meta-analysis of the outcomes of structured telephone support or telemonitoring as the primary component of chronic heart failure management in 8323 patients: Abridged Cochrane Review. Eur. J. Heart Fail. 13, 1028−1040.

Kamel, A., Tawfik, B., 2010. Decision Support Systems in clinical engineering, In: 2010 5th Cairo International Biomedical Engineering Conference. pp. 197−201.

Lavy, S., Shohet, I.M., 2007. Computer-aided healthcare facility management. J. Comput. Civ. Eng. 21, 363−372.

Luschi, A., Marzi, L., Miniati, R., Iadanza, E., 2014. A custom decision-support information system for structural and technological analysis in healthcare. XIII Mediterranean Conference on Medical and Biological Engineering and Computing 2013. Springer, pp. 1350−1353.

Luschi, A., Miniati, R., Iadanza, E., 2015. A web based integrated healthcare facility management system. 6th European Conference of the International Federation for Medical and Biological Engineering. Springer International Publishing, pp. 633−636.

Miniati, R., Dori, F., Iadanza, E., Fregonara, M.M., Gentili, G.B., 2010. Health technology management: a database analysis as support of technology managers in hospitals. Technol. Health Care. 19, 445−454.

Rajalakshmi, K., Mohan, S., Babu, S., 2011. Decision support system in healthcare industry. Int. J. Comput. Appl. 26, 42−44.

Rodriguez, E., Miguel, A., Sanchez, M.C., Tolkmitt, F., Pozo, E., 2003. A new proposal of quality indicators for clinical engineering. Engineering in Medicine and Biology Society, 2003. Proceedings of the 25th Annual International Conference of the IEEE. IEEE, pp. 3598−3601.

Shim, J., Warkentin, M., Courtney, J., 2002. Past, present, and future of decision support technology. Decis. Support Technol. 33, 111−126.

Takeda, A., Taylor, S.J.C., Taylor, R.S., Khan, F., Krum, H., Underwood, M., 2012. Clinical Service Organisation For Heart Failure. The Cochrane Library.

Wagner, E.H., Glasgow, R.E., Davis, C., Bonomi, A.E., Provost, L., McCulloch, D., et al., 2001. Quality improvement in chronic illness care: a collaborative approach. Joint Comm. J. Qual. Patient Saf. 27, 63−80.

Yancy, C.W., Jessup, M., Bozkurt, B., Butler, J., Casey, D.E., Drazner, M.H., et al., 2013. 2013 ACCF/AHA guideline for the management of heart failure: a report of the American College of Cardiology Foundation/American Heart Association Task Force on Practice Guidelines. J. Am. Coll. Cardiol. 62, e147−e239.

8

Early Stage Healthcare Technology Assessment

Leandro Pecchia[1,2], Rossana Castaldo[1], Paolo Melillo[3], Umberto Bracale[4], Michael Craven[5], Marcello Bracale[4]

[1]SCHOOL OF ENGINEERING, UNIVERSITY OF WARWICK, UK [2]HEALTHCARE TECHNOLOGY ASSESSMENT DIVISION, INTERNATIONAL FEDERATION FOR MEDICAL AND BIOLOGICAL ENGINEERING (IFMBE), PARIS, FRANCE [3]DIPARTIMENTO DI OFTALMOLOGIA, SECONDA UNIVERSITÀ DEGLI STUDI DI NAPOLI, NAPLES, ITALY [4]DEPARTMENT OF SURGICAL SPECIALITIES, NEPHROLOGY UNIVERSITÀ DEGLI STUDI DI NAPOLI FEDERICO II, NAPOLI, ITALY [5]UNIVERSITY OF NOTTINGHAM, NOTTINGHAM, UK

Healthcare Technology

The term "Healthcare Technology" is used to describe any intervention that may be used for safe and effective prevention, diagnosis, treatment, and rehabilitation of illness and disease, for instance:

1. drugs (from an engineering perspective, this relates to drug delivery, enabling control of the drug release, absorption, distribution, and elimination rates);
2. medical devices, as defined by the European Directive 93/42/EEC (medical devices in general)
3. procedures (e.g., surgical techniques, acupuncture, medical advice);
4. secondary care (e.g., hospitals, outpatient);
5. health programs (e.g., screening programs, public health).

Health Technology Assessment (HTA) is a scientific, multidisciplinary, and multidimensional decision-making process that enables benchmarking of the positive and negative effects of such technologies. The purpose of HTA is to holistically consider a variety of alternatives and select the best technology that addresses the identified medical (and nonmedical) needs in order to support health policy decisions at all levels.

However, satisfying such needs in the healthcare domain is quite challenging since improving or maintaining a health state depends only partly on technology action. Furthermore, the technology contribution itself is also often complex and may be difficult to measure and evaluate. A widely deployed scheme that summarizes the production process of National Health Services (NHS) and other concomitant factors that affect people's health is shown in Figure 8.1.

Clinical Engineering. DOI: http://dx.doi.org/10.1016/B978-0-12-803767-6.00008-8

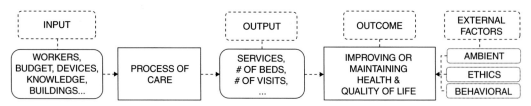

FIGURE 8.1 Production process in health.

In accordance with Figure 8.1, it is worth considering three concepts (Anthony and Young, 2003):

1. $Efficacy = \dfrac{Output}{Input}$

2. $Efficiency = \dfrac{Outcome}{Output}$

3. $Performance = Efficacy * Efficiency = \dfrac{Outcome}{Input}$

Taking in account those three concepts introduced using a system engineering approach, we can say that the purpose of HTA is to identify the technology with the best performance. HTA is thus the decision-making process through which a decision can be made whether to adopt a new technology in health care, based on its efficacy and efficiency with respect to alternatives (Craven, 2007).

In order to maximize its impact, it is important to consider such variables early on, before a new biomedical technology arrives on the market and, where possible, during the research and development phase of the technology. The World Health Organization (WHO) emphasizes that technology assessment may be perceived as an obstacle that slows down the introduction of innovative technologies in health systems; however, one of the reasons identified by the WHO is that the "assessment problem" is considered only when the technology is entering the market (Health technology assessment of medical devices, 2011). The growing importance of HTA for the biomedical engineering profession is increasingly clear. For example, European projects aiming to train biomedical engineers (CRH-BME Curricula Reformation, 2008) have for the first time (in 2010) included HTA in teaching foundation for biomedical engineering courses (Jarm et al., 2012; Pallikarakis et al., 2011). Furthermore, on a global scale, the International Federation of Medical Engineering and Biomedical (IFMBE) created the HTA Division, to minimize this problem (IFMBE, 2012).

Health Technology Assessment: Standard *De Facto*

Depending on the evaluation purpose and the available resources and time, HTA can take various formats. However, HTA reports published in respected international scientific journals generally have a very regular structure, which defines a *de facto* standard of methods and tools for HTA. The basic structure of many HTA reports can be summarized as:

1. definition of the medical goal and decision problem;
2. assessment of consequences using clinical evidence;

3. resource assessment using cost analysis;
4. analysis of incremental cost versus consequences.

In literature, several algorithms summarize the main steps of an HTA. Figure 8.2 shows a general diagram along with a brief description of the individual steps (Pecchia et al., 2009a,b,c).

Definition of the Medical Goal and Decision Problem

Different scenarios can affect the definition of the medical problem, which is often determined by contingencies depending on the size of the problem itself. Therefore, it is not always easy to identify a standard method for the choice of the medical problem. Although many authors suggest assessing the consequences of a health technology in several dimensions (economical, technical, ethical, etc.), the majority of studies are focused mainly on the clinical and the economical dimensions. While this is in theory a limitation, these two dimensions are sufficient for the majority of HTA studies. Later on we will consider other dimensions of HTA.

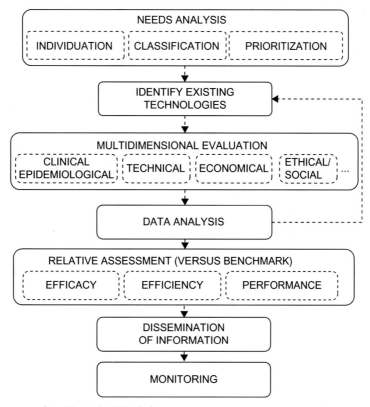

FIGURE 8.2 HTA process. *(Pecchia et al., 2009a,b,c).*

Needs Analysis

In order to carry out an HTA, the first step is to formally define the medical problem (usually clinical or epidemiological) that the technology in question must meet, through a detailed analysis of needs. In this case, the analysis includes: identification of needs; their organization into classes (e.g., clinical, economic, technological); the attribution to each need and each class a weight that reflects its relative importance. In other words, the relevance and importance of each identified need must be quantified. The quantification process is crucial for selecting the best performing technology and should be completely transparent and structured (weighted) *a priori*. Moreover, since the quantification process is based on underlying qualitative information, it should be carried out as objectively as possible in order to make the evaluation reproducible (Pecchia et al., 2011a,b). Several methods have been used in the literature for scientific needs assessment, such as *Conjoint Analysis* (CA) (Bridges et al., 2012), *discrete choice experiments* (de Bekker-Grob et al., 2012), *best—worst scaling* (Gallego et al., 2011) and *Analytic Hierarchy Process* (AHP) (Pecchia et al., 2011a,b).

At the end of this step, the most important needs to be met should be known and the most significant dimensions of the analysis (e.g., clinical/epidemiological, economic, technical) should be identified and quantified with their relative importance.

Technology Identification

In this step, potential technologies that can help meet the needs are identified. It is good practice at this stage to not consider *a priori* partial information on technology performance. However, the major difficulty is to include all the technologies that might contribute to meet the identified needs. For example, technology identification is undoubtedly affected by the selected assessment scale, which differs between national, regional, or institutional levels, and according to the scale chosen, there are different HTAs: macro-, meso-, and micro-HTA. For example, a medical device assessment for a health institution (e.g., hospital) may be mainly aimed at optimizing the purchase of new equipment to fulfill a specific treatment objective in that institution alone, whereas an assessment at a regional or national level may have as its ultimate objective the optimization of strategic health outcome within the territory, which would broaden the range of technologies to be considered.

Resources and Consequences Evaluation

Resources and consequences are generally considered as the two fundamental dimensions of HTA considering both economic cost and clinical/epidemiological outcome (including quality of life or other quantified benefits). In the following section, we also consider ethical/social and the technical dimensions of HTA.

The Economic (Cost) Dimension

Along with the clinical dimension, the economic (cost) dimension is most often considered in HTA processes. In fact, some HTA processes are limited to cost alone (cost studies).

At this step of the analysis, the costs of all the economic resources used for the provision of technologies, taken in consideration, need to be identified.

Without going into details of specific methods, the purpose of this step is to assign an economic value to each resource used directly (or in some analyses, indirectly) in the production of the provided health service. By resources, it is meant: human resource complexity, environment, technologies, supplies, depreciation, cleaning, insurance, and so forth.

In this chapter, we mainly consider direct costs which are the costs of delivering the treatment by the institution, health service, or insurer (payer's costs) and do not include indirect costs, e.g., travel costs to individuals or mainly productivity losses to society caused by the health problem or disease (although see later with respect to cost−benefit analysis which may include consideration of indirect monetary benefits). Within direct costs, it is useful to distinguish between fixed cost and variable cost. Fixed costs are costs that are independent of the amount or number of good/service provided whereas the variable costs are costs that vary with it. For example, cost of buildings rent is generally independent of the number of services provided which is a fixed cost, whereas a surgical procedure may incur a variable cost due to staff wages, facility time, and consumables used each time, which are carried out. Often, capital equipment can be considered as a variable operating cost, divided into the number of procedures per year.

Economic (cost) analysis is divided in three steps: identification of necessary resources, the classification of the same, and monetization of their value. Further complexity is introduced when the analysis is carried out over a longer period (time horizon); in which case, discounting is generally used to account for the value of money in subsequent years, and price or currency changes may also need to be modeled.

The Clinical/Epidemiological Outcomes Dimension

This is another crucial step for HTA. In this step, the technology clinical/epidemiological effects are evaluated through the outcome measurement. Of course, the outcome varies according to various factors as before, including: the assessment scale (e.g., institutional, regional, national), the type of technology that is being analyzed (e.g., equipment, medication, health process, surgical technique), the purpose of the technology (e.g., diagnosis, care, treatment, prevention, screening), and the type of patients concerned (e.g., age, medical specialty). It is very common to classify a study according to the method chosen to measure the technology effect. Classic examples of measurement of technologies effects are effectiveness, utility, and benefits estimation.

By effectiveness, it is meant the measure of how technology meets a specific clinical outcome (e.g., reduced mortality, pain, or maintenance/improvement of a "natural" biomedical parameter such as blood pressure). The main limitation of this analysis is that it may measure outcome in a one-dimensional manner, although several one-dimensional outcomes can be combined, with the added difficulty of deciding which outcomes are most important. In general, considering costs against such outcomes is called a cost-effectiveness analysis (CEA), where the appropriate parameters of effectiveness are specified. Sometimes, instead of attempting a weighting, the result may be presented to decision-maker as a side-by-side table of costs and outcomes in the HTA report, in which case it is called a cost-consequences analysis (CCA). See later for a summary of the different types of analysis. Utility is a measure of health state, considering simultaneously multidimensional functions of several clinical outcomes, used to estimate the quality of life. It is common to measure the effect of a technology-supported treatment by its contribution to improvement in quality of life. Here the outcome is estimated as a function independent dimension as: mobility, self-care, usual activities, pain/discomfort, and anxiety/depression; the absence of pain. A common example of the utility function is the one used by QALY (Quality Adjusted Life Years), which weights the years of life gained with a quality index that takes value 1 for a year of life in perfect health, and a value between 0 and 1 for years with less quality. These values are then added together in order to assess the overall value of the QALY in the observation period. As an example, this method equates two years lived at 50% of perfect health with one year lived in perfect health. This is a somewhat indirect measure because it estimates the global health state based on preferences gained from interviews with people in the wider society (not only patients who have the disease). The main methods are:

1. Time trade-off (TTO), interviewees choose between being ill for a number of years or be healed and restored to a healthy condition, but having a minor life expectancy;
2. The standard gamble, in which interviewees choose between remaining in the current conditions or to undergo treatment, through which the patient is likely to die or be restored to good health;
3. Rating scales, such as the visual analogue scale (VAS), where interviewees are asked to assign a value between 0 (death) to 100 (health) to the different health states.

The major limitation of this method is in the way a specific status is defined. This assessment is often subjective and difficult to generalize. There are also other structured methods for self-assessment of health status. Unfortunately, no scale proposed so far has reached consensus and is thus universal which is a limitation of the use of utilities and QALYs, although there are agreed data sets used in some regions such as the EQ-5D value sets which are based on TTO and VAS (Oppe et al., 2007). The type of analysis comparing costs and QALYs is called a cost-utility analysis (CUA).

Another way to address the problem is by a cost–benefit analysis (CBA). Benefit is the effect of a health outcome expressed in monetary terms (we note, it is common to hear the term "benefit" used to express the more general concept of clinical outcome, generating some confusion). Monetary quantification of the value of health can be carried out with the two widely used techniques:

1. Human Capital, HC;
2. Willingness To Pay, WTP.

The HC standard method estimates the value of an individual's life through their future potential productivity to society, calculated as the value of the remuneration of the planned work, reported at present value. This method is implicitly based on the maximization of such productivity as the goal. The main problem with this approach is the reduction of the individual health to only include the value that a person provides to Gross Domestic Product (GDP), neglecting the intangible value of life itself. In fact, in a country with a public NHS and with a constitution that emphasizes the equality of all citizens, regardless of how much they contribute to GDP, which is still a questionable indicator of the wealth of a country, this method is still not favored in many countries.

The WTP standard method estimates, in the case of a pathology, how much a society would be willing to pay to avoid contracting the disease. In other words, it estimates the aggregate value that a population at risk would invest on programs that save lives statistically, or the sum of the amounts that individuals would be willing to pay *ex ante* to reduce the probability of their death. An example may help to illustrate this point. Suppose that every person in a population of 100,000 inhabitants is willing to pay 25 pounds ($25 \times 100,000 = 2.5$ pounds) for a program that is expected to reduce the overall probability of death from 0.09% to 0.08%. Since this is equivalent to a reduction in the mortality rate from 90 per 100,000 to 80 per 100,000, the implied value for each of the 10 lives saved is 250,000 pounds (2.5 million divided by 10 lives saved). Of course, whether a life is worth more or less than this figure is a difficult question for society.

The Ethical and Social Dimension

The consideration of these dimensions in the HTA reports is very limited (Sacchini et al., 2009). For more information, please refer to specific texts (Reiser, 1988). However, it is useful to recall the pattern of Heitman (1998) suggesting to organize ethical dimension in five categories:

1. aspects of concepts and definitions of the process of HTA itself;
2. aspects related to diagnostic procedures (e.g., limits, risks, employment.);
3. aspects related to preventive strategies (e.g., the management of the risk of pathologies) and therapies (e.g., evidence, efficiency, appropriateness);
4. aspects related to research (e.g., protection of subjects enrolled in the studies, informed consent);
5. aspects related to the resource allocation (e.g., distributive justice, rationing mechanisms, economic evaluations).

The Technical Dimension

As far as the technical dimension is concerned, this is related predominantly to technological aspects, which affect only partially or indirectly the economic or clinical dimensions. It includes, for example, aspects related to the drug administration that do not impact directly on the patient (preparation, archiving, etc.). Moreover, the technological asset of an institution affects the technical dimension. For example a hospital linked to a university which develops technologies may be more comfortable with the deployment of new technologies such as robotic surgery or may be able to provide it at minimal cost, for example, if linked to research activities funded externally, making its introduction more attractive.

Data Analysis: The Meta-Analysis

Data analysis presents the next step. This includes data collection, processing, synthesis, and representation. The most common method used to collect clinical data for HTA is from meta-analysis (Sutton, 2000). This method uses statistical tools for harmonizing and combining results from different study to identify significant patterns. Therefore, the main aim is to aggregate information in order to achieve a higher statistical power for the measure of interest, as opposed to a less precise measure derived from a single study. Although this analysis has several limitations, it enables to conduct studies on large number of patients and on different case studies, decreasing for instance the economic concern. The studies' identification is usually performed by independent researchers through structured approaches. Subsequently, their results are compared in order to reach consensus on what studies accepting to the next step of the analysis, namely, the elaboration of the latter. The inclusion criteria, used to define which studies are worth elaborating, are defined *a priori*, by choosing the appropriate criteria, including the choice of scientific studies with a high evidence level. For instance, the randomized controlled trials (RCTs) are included as first, pseudo-randomized trials as second, and finally the cohort studies (prospective, prospective with retrospective control group, retrospective) or the retrospective studies (Guyatt et al., 1995). After having identified the studies, the analysis proceeds with the classification and weight of the studies in order to extrapolate the results. Finally, the results of individual studies are aggregated, weighing them on the basis of the study importance. This importance depends on the number of cases handled and the accuracy of the results. These results are then presented in graphic form, to be communicated to the larger number of decision-makers, which may not be very confident with advanced mathematical methods. Figure 8.3 shows the results of a meta-analysis carried out on a biological drug to evaluate its effects on life quality between patients assuming the drug (treatment group) and control group.

This meta-analysis was conducted on five studies (one per line) (Furst et al., 2003; Van de Putte et al., 2003; Weinblatt et al., 2003; Keystone et al., 2004; Van de Putte et al., 2004). In each study, the effects were observed in a treatment group, with a sample of N_t subjects

Study	Treated group N_t mean (SD)		Control group N_C mean (SD)		WMD (fixed) (95% CI)	Weight (%)	WMD (fixed) (95% CI)
van de Putte, 2003	71	−0.45 (0.46)	70	−0.04 (0.37)	■	11.40	−0.41 (−0.55 to −0.27)
van de Putte, 2004	225	−0.38 (0.61)	110	−0.07 (0.49)	■	14.67	−0.31 (−0.43 to −0.19)
STAR	312	−0.51 (0.56)	314	−0.26 (0.48)	■	32.37	−0.25 (−0.33 to −0.17)
ARMADA	67	−0.62 (0.63)	62	−0.27 (0.57)	■	5.04	−0.35 (−0.56 to −0.14)
Keystone, 2004	419	−0.60 (0.56)	200	−0.25 (0.56)	■	24.30	−0.35 (−0.44 to −0.26)
Total (95% CI)	1094		756		◆	87.78	−0.31 (−0.36 to −0.26)

Test for heterogeneity: $x^2 = 4.90$. df = 4 ($p = 0.30$). $I^2 = 18.4\%$
Test for overall effect: $z = 12.41$ ($p < 0.00001$)

$$-1 \quad -0.5 \quad 0 \quad 0.5 \quad 1$$

FIGURE 8.3 Mean difference of adalimumab effect on HAQ, in the control group and the treatment group. In the first five rows, are reported the results of other studies. The end result is shown in the bottom line.

(from 67 to 419 per study), and compared with a control group of N_c subjects (from 62 to 314 per study). The quality life of a subject has been measured in the individual studies, by the Health Assessment Questionnaire (HAQ), which measures the outcome for clinical trials, and it is widely used before and after treatment in the two groups (treatment and control group). HAQ values, indicating an improvement of health status, are shown in the second and fourth columns, respectively, for the treatment and control groups, respectively; the mean and standard deviation values are reported in parentheses. In the penultimate column, the relative weight of each study is shown. The weight increases with the number of subject of the two groups (N_c and N_t) and decreases with the magnitude of the standard deviation (SD) of the effect. In fact, the study of Keystone (fifth line) conducted on 619 subjects (419 + 200), and with an SD of 0.56 in both control and treated group, has a relative weight, normalized to percent of 24.30, which is less than the relative weight of the STAR study, conducted on a larger number of subject but with a less SD value, at least in the control group. For instance, the STAR recruited 626 patients (312 + 314) and reported an SD of 0.56 in the treated group, but an SD of 0.48 in the control group. Finally, the last column shows, for each study, the mean difference between groups (control against treated) and in brackets the confidence interval of 95%. In the last row, Weighted Mean Difference (WMD) is presented; it shows the difference that would have been achieved if the study had been conducted only one study of 1094 patients treated with the drug and 756 patients for the control group. Similar statistics on dichotomous variables can be found in Sutton (Sutton, 2000). Many examples of meta-analysis can be found in the existing literature (Castaldo et al., 2015; Bracale et al., 2012a,b). The second of the last two studies (Bracale et al., 2012b) is a network meta-analysis, which represents one of the possible generalizations of the meta-analysis (Lumley, 2002).

Cost and Consequences Assessment

In practice, it never happens that a new technology overcomes the other existing in all the dimensions of the analysis. Therefore, it is crucial to understand which dimensions need to be prioritized, according to the objective of the analysis. This decision may depend strongly on the analysis scale. For example, a Local Health Authority may consider as priority the maximization of the efficiency, while in a highly specialized hospital it may consider as priority the maximization of the efficacy, and in particular certain outcomes. As discussed in previous sections, the majority of studies consider the economic and clinical dimensions in terms of efficacy, utility, or benefit. In the literature (Nord, 1999), four methods to analyze the efficacy of a health technology are Cost Minimization Analysis (CMA); Cost-Effectiveness Analysis (CEA), which may also include Cost-Consequences Analysis (CCA); Cost-Utility Analysis (CUA); and Cost–Benefit Analysis (CBA). Table 8.1 summarizes the main characteristics, advantages, and limitations of these four methods. In addition, in Figure 8.4 the algorithm, to choose the most suitable type of analysis, is described.

There may be three conditions, as summarized in Table 8.2 (in which the efficacy is treated in the same way as the utility), during the evaluation of the costs and the outcomes/consequences of a new technology compared to a widely used technology.

Table 8.1 Comparison of Different Methods Used to Analyze the Cost/Result

	CMA	CEA	CUA	CBA
Costs	Monetary units	Monetary units	Monetary units	Monetary units
Consequences	Equal in both programs	Clinical outcomes	QALY	Monetary units
Measuring	Differences in costs (DC)	ICER	ICUR	ICBR
Advantage	Direct measurement	Direct measurement	Indirect measurements	Indirect measurements
	Necessary for the other	Uniform clinical outcomes	Mixed clinical outcomes Multidimensional analysis	Mixed outcomes Multidimensional analysis
Limits	No consequences	One-dimensional analysis	Indirect measurement	Indirect measurement
		Data table missing in many national health services	Data table missing in many national health services	Monetization of value of life Ethical limits

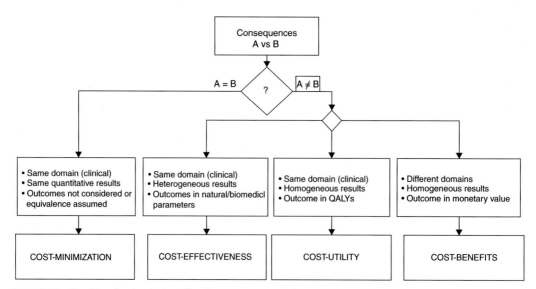

FIGURE 8.4 Algorithm for the choice of health economic analysis.

If a new technology (Technology A) proves to be less effective and more expensive compared to a benchmark (gold standard—technology B), as represented in the first case of Table 8.2, it is rejected without further analysis, unless there is a high degree of uncertainty in the data (in this case, further experiments are needed). On the other hand, a technology is a good candidate, if it is more effective and less costly (the second case of Table 8.2). In the third case, the achievement of a greater efficacy is related to an additional cost, therefore the

Table 8.2 Possible Results of a Cost-Effectiveness Analysis

Costs	Consequences	Result
$C_A \geq C_B$	$X_A < X_B$	DOMINANCE. A is more expensive and less effective/utile than B. B continues to be the standard.
$C_A \leq C_B$	$X_A > X_B$	A IS COST SAVING. A is less expensive and more effective than B. The new technology A is introduced gradually in the NHS.
$C_A \geq C_B$	$X_A > X_B$	INCREMENTAL ANALYSIS IS REQUIRED. A is more expensive and more effective/utile than B. The new technology A may be introduced gradually in the NHS if the cost for unit of effectiveness/utility is less than the last one introduced into the NHS.

technology requires an evaluation of the Incremental Cost-Effectiveness (or Utility) Ratio (respectively ICER or ICUR) expressing the cost of a unit improvement in outcome. This third case is a common situation; since the price of a new technology includes the manufacturer's recouping of design and prototyping costs, they can be much higher (especially with low volume production of devices) than a product already on the market. An ICER over a given threshold may suggest that the new technology is more cost-effective, and therefore its adoption is justified since the expected outcomes are considered to be worth the costs incurred. In the United Kingdom, it is generally considered that for the adoption of a new drug or device-related treatment, a threshold of between £20 K and 30 k pounds per QALY gained is acceptable. In other words, it is considered worth spending up to 30 k pounds for 1 year of life gained or for 2 years of life lived to at 50% of perfect health compared with no intervention or intervening with a benchmark treatment.

Standard Method Limitations and Weaknesses

Most HTA methods, which have been effectively used for evaluation of drugs, may straggle when applied to the assessment of medical devices. In fact, there are some differences between drugs and medical devices, which have a significant impact on the assessment (Table 8.3).

Furthermore, HTA methods are not very accommodating of the reality of research and development process for medical devices. For example, in the research phase, there are generally no RCTs that are sufficiently large or numerous for a proper assessment of the consequences. Nevertheless, it is important to aim to collect information for HTA at a very early stage of research, preferably during the conception of the idea, together with a reliable estimation of the probability of return on investment or of the needs of the market to which the technology aims to (Ijzerman and Steuten, 2011).

Finally, standard HTA methods do not directly identify the priorities among the needs of individual users or subgroups. For example, the weighting of utility in value sets is not

Table 8.3 Comparison between Drugs and Medical Devices

Devices	Drug
Principal action	
Other than principally drugs	Pharmaco/Immunologic/Metabolic
Mechanical/Electromagnetic/Materials	Chemical based
Product life cycle	
Short life cycle	Long life cycle
Constantly evolving components/parts	Unchanging compound
Clinical evaluation	
Difficult to blind (no placebo)	Easy to blind
Multiple end users	Usually one end users
Long learning curve	Short learning curve
Strongly dependent by settings/users	Less dependent by settings/users
Complex to standardize for RCT	Easy to standardize for RCT
Use issues	
User-dependent efficacy	Efficacy is less user-dependent
Often require intensive training	Usually do not require training
Complication decrease with use	Complication increase with use
Diversity	
Mainly small companies/few large co.	Mainly large multinationals
Diagnostic or therapeutic	Therapeutic
Costs	
Varying overheads/slow return	High overheads with quicker return
Higher distribution costs	Lower distribution costs
Higher maintenance/installation costs	No maintenance/installation

(Adapted from Craven, 2007).

reflective of age so dimensions pain and mobility are averaged in one country and are the same for people with different aspirations. Therefore, stratification by age and other factor is possible in principle (Kind et al., 1998).

Early Stage HTA (eHTA)

Recent studies try to overcome some of the limitations described in particular: does the assumption that the device, on which the research is to be carried on, work exactly as demonstrated by the experimental data and provide benefits or cost savings (or both) more than the existing technology (and is therefore reimbursable by the NHS)? This is exactly the question that the early stage HTA tries to answer scientifically. In this section, some of these methods are introduced.

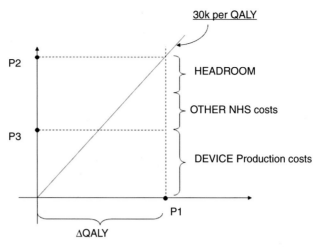

FIGURE 8.5 Headroom method.

Headroom Analysis

In order to answer the aforementioned question in an appropriate way, outcomes should be quantified and expressed in terms of QALYs and compared to a reference threshold for the incumbent technology whether to be considered cost-effective, as discussed previously. As stated previously the NHS generally considers appropriate a reimbursement of up to £ 30,000 per QALY gained, for one patient undergoing a novel treatment versus an existing alternative (which may include no treatment). The Headroom method tries to answer the above question on the supply side (i.e., from the manufacturer's perspective) (Cosh et al., 2007) by evaluating the technology's cost-effectiveness through the ICUR, according to the best hypothesis. Therefore, since the evaluation is carried out, according to the most favorable assumptions, the results do not guarantee that the device will be successful, but it may exclude ideas being not successful. The method consists of the following steps (Figure 8.5):

1. assume that the device actually improves the estimated ΔQALY (Step 1);
2. calculate the maximum cost that the NHS is willing to spend on such ΔQALY (Step 2);
3. subtract the cost of the device production (Step 3);
4. subtract the costs that the device does not reduce (e.g., administration);
5. calculate the maximum profit margin for the product.

If the final gain presents a sufficient margin, the R&D could reasonably continue. Otherwise further research investment would be better spent elsewhere. Having this knowledge at a very early stage in the applied research is very useful.

Cost-Utility Evaluation in a Preliminary Phase: Markov Models

As introduced in previous sections, cost analysis assigns a monetary value to each resource used in the care process. This analysis can become more complex when the needed resources for each patient vary according to disease evolution (for instance, deterioration may require additional resources). At the beginning of an analysis of a typical long-term or chronic health condition, patients in the same study population may be in different early stages of the disease. Subsequently, each state may vary: the probability of having an exacerbation, severity of exacerbation (mild, moderate and severe), the probability of moving from one state to another (e.g., mild becoming moderate). Such phenomenon can be modeled using a Markov model (Sonnenberg and Beck, 1993). These models assume that a patient is always attributable to one of a finite number of discrete health states called Markov states (nodes into the model). All the events are represented as transitions from one state to another (edge), with a given probability "p." A marginal cost "ΔC" is associated to each event (transition) "k", representing the total amount of required resources for that event, including costs for normal treatment, treatment of exacerbations, and so forth.

Since some consequences can vary over a time period, Markov models can be used to perform an incremental health economic analysis (Sonnenberg and Beck, 1993) in the evolving patient population.

In this case, the following steps are usually performed:

1. Each status "k" of a pathology (including the initial one) is associated to a cost (ΔC_k), which represents the total amount of required resources such as trading patients in such status, and a marginal consequence, which may be a marginal effectiveness variation (ΔE_k) or a marginal utility (ΔU_k).
2. The transition from one state to another is associated with a probability.
3. The expected exacerbations from each state are associated with a probability, for each kind of exacerbation (e.g., mild or severe).
4. The total path costs for each final state and the total path effectiveness (or utility) are combined representing the total costs (C) and total consequences (E or U) for groups of patients in each final state.
5. The model is then evaluated dynamically (often with a Monte Carlo simulation).

This generates a distribution of Cost-Utility points (C, U) j, with $j = 1, \ldots, N$ (number of simulations carried out), as shown in Figure 8.6.

In the case of a comparison between a new technology (A) and a benchmark (B), this procedure is repeated with the same Markov model, but with probability, cost and consequences resulting from the use of the new technology in order to calculate the costs and consequences of each one ($\Delta = A - B$). In this case, the cost-consequences plane has on the x-axis the consequences difference (ΔE or ΔU) and on the y-axis the costs difference (ΔC) (Chapman et al., 2000).

In the development of a new device or a technological breakthrough, it can be difficult to obtain data that enables stratification of patients according to risk factors or previous interventions (if any), or to consider the whole range of outcomes that may occur. However, it may be sufficient in a preliminary stage to consider the main outcomes and to limit the patient stratification to a small number of sub-groups. One of the most common analytical approaches is based on the use of Markov models with a small number of states and

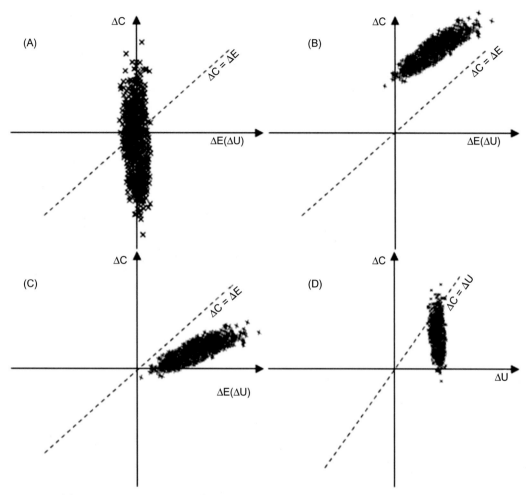

FIGURE 8.6 (A) Equivalence between the efficacy of two technologies (symmetry with respect to the vertical axis); (B) new technology less cost-effective than the benchmark; (C) new technology more cost-effective compared to the benchmark as all points of the simulation are below the line of willingness to pay; (D) uncertainty: the new technology seems to be more effective in the 80% of the simulation.

transitions according to the availability of data. This approach was applied by Dong and Buxton (Dong and Buxton, 2006) to compare a new computer-assisted technology and the standard procedure for total knee replacement (TKR). The number of patients who had a primary TKR was limited to three groups, according to risk: the absence of complications, minor complications, and serious complications. Patients with complications may require revision surgery or other treatments. The model required a number of different variables, such as the transition probabilities between states, the various additional complications or death, which were extracted from the literature or estimated ad hoc. The comparison was made with a model with nine states, proving that the assessment in the early stage offered

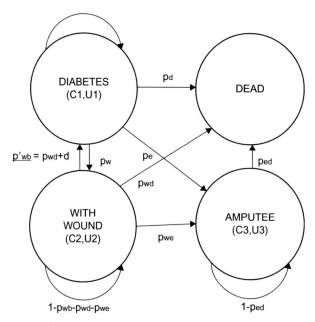

FIGURE 8.7 Evaluation model in the early stage via Markov model.

sufficient knowledge of effectiveness long before an RCT or meta-analysis were available in the literature. In a subsequent study (Craven and Morgan, 2011), a model with four states was used for what-if analysis regarding a device promoting healing of diabetic foot ulcers. The model included four states: absence of injury, the wound (ulcer), amputation, and death. This is a simplification of a more sophisticated model that included classification in high and low risk and considered more types ulcerations and amputations with different probabilities. The value of the device was examined by assuming that the probability of transition between two states (return from ulcerated foot to the absence of wound) was improved by the use of the device in question, as illustrated in Figure 8.7.

The Needs Analysis and Benchmarking via AHP

As introduced previously, several methods have been proposed for the analysis of the HTA needs, including Conjoint Analysis (CA) (Bridges et al., 2012), discrete choice experiments (de Bekker-Grob et al., 2012), best—worst scaling (Gallego et al., 2011), and recently the Analytic Hierarchy Process (AHP) (Pecchia et al., 2011a,b). According to (Scholl et al., 2005), the AHP method proved to be more suitable than the CA for complex decisions involving many factors. Mulye (1998) has suggested that AHP is more effective than CA, since more than six attributes are considered. Ijzerman (Ijzerman et al., 2008) concluded that AHP, compared with CA, is shorter, more flexible, and easier to implement in questionnaires, although inconsistencies can be generated and other methods may benefit of a more holistic approach. This section briefly describes the method and its application to eHTA.

AHP is a decision-making method aiming to solve multifactorial and multidimensional problems. This method is particularly effective in quantifying the users' opinions based on their personal experiences and in taking a coherent decision. AHP is based on the classification of a hierarchy of elements; therefore, questionnaires are showed at each user, which, through pairwise comparisons, assess the relative importance of the elements. By asking redundant questions, it is possible to assess the consistency of the respondent. Applying this iterative method, it is possible to estimate: the relative importance of each need in its category, the relative importance of each category, and the absolute importance of each need compared to all the other. A detailed explanation of the AHP method applied to user need elicitation can be found in Pecchia et al. (2009a,b,c) and Pecchia et al. (2011a,b). This method has been successfully applied to detect the user needs in previous studies, to identify the best care model for chronic heart failure (Pecchia et al., 2009a,b,c) for the choice of a maintenance contract based on the location of the hospital and the services offered (Pecchia et al., 2009a,b,c) for the identification of risk factors for falls in the elderly (Pecchia et al., 2011a,b), or the purchase of a computed tomography scanner (Pecchia et al., 2013) and for the definition of a life quality scale for the well-being of the elderly (Pecchia et al., 2011a,b).

The method consists of five steps:

1. identification of the needs that new biomedical devices aim to satisfy;
2. design of a need tree with nodes (categories) and leaves (needs);
3. development and presentation of a questionnaire to detect users' needs;
4. weights quantification; and
5. assessment of multiple technologies.

The needs' identification (first step) is generally conducted revising the scientific literature and through focus groups, involving domain experts (e.g., medical specialists, clinicians, biomedical engineers), potential users of the device (medical personnel, patients), and one AHP expert. In the second step, the needs, identified previously, are reorganized in a tree structure. Three main categories are included in most tree models: clinical needs, economic needs, and technical needs (Pecchia et al., 2009a,b,c).

The third step is based on the development of questionnaires, by which respondents compare the relative importance of each need against the others of the same category. For example, for each pair of needs (i, j), each respondent is asked: "Based on your experience and according to the specific evaluation objective, how do you rate the need i against the need j?" The expert chooses one of the suggested choices, relating to the question. A weight is then assigned to each judgment using a numerical scale. Several scales have been proposed, including the natural scale, proposed by the same inventor of the method Prof. Saaty (Saaty, 1977), which consists in assigning a numeric value to positive choice, e.g., from 3 "more important" to 5 "very more important". A value of 1 is assigned to the choice "equally important." The reciprocal values are attributed to the remaining choices: for 1/3 "less important", 1/5 if "much less important." The natural scale of Saaty provides nine possible choices, (1, equally important; 9, extremely important) (Saaty and Vargas, 2001). The fourth step involves a series of numerical calculations aiming to find the relative weight, reflecting the relative importance, of each element (need or category), compared with all the other in the same father node. Details can be found elsewhere (Pecchia et al., 2010; Saaty, 1982).

Finally, the same algorithm is applied to compare device under development with existing technologies (Fifth step). In other words, the question posed this time is, "how much the technology A, in comparison with the technology B, is meeting the need i?" Using the same weights as described above, it is possible to obtain a vector $\underline{E}$ (with m elements, where m is the total number of identified needs) that describes the efficacy of the technology in the development phase, compared to all others. The summary analysis that represents the overall performance of the device in question compared to other existing technologies is estimated by calculating the scalar product of this vector with the vector GW introduced earlier. Examples of such applications applied to technologies may be found in the literature in various stages of development (Pecchia et al., 2009a,b,c; Ijzerman and Steuten, 2011; Pecchia et al., 2007; Liberatore and Nydick, 2008; Sloane et al., 2003).

Conclusion

In this chapter, after briefly recalling some key concepts such as efficacy, efficiency, and health performance, some HTA methods were introduced. These methods are very useful and their widespread distribution is highly desirable. However, they have historically been developed to evaluate drugs and have limitations of applicability when they are used to evaluate new medical technologies, particularly to estimate the potential for sustainability in development, research, or even conception (early stage).

In the last few years, evaluation methods designed precisely for the early stage of HTA biomedical technologies have been developed; they aimed to answer the following question: "Assuming that the device on which you are carrying out the research work exactly as demonstrated by experimental data, which benefit will this technology give to the patient more than the technologies currently on the market ?" In this chapter, three of these methods were presented. In this period, assessments are particularly difficult for several reasons, including the nonavailability of adequate data. However, it would be helpful to have an assessment, albeit preliminary, to decide whether and how to invest resources. In fact, in the research and development process, great results in terms of sustainability of the technology itself are obtained with small changes. In recent years, the biomedical international community has begun to show more interest in HTA, due to applicability of these methods also to research. Moreover, the use of these methods may speed up the commercialization of devices.

References

Anthony, R.N., Young, D.W., 2003. Management Control in Nonprofit Organizations. 7th ed. McGraw-Hill/Irwin, Boston.

Bracale, U., Melillo, P., Pignata, G., Di Salvo, E., Rovani, M., Merola, G., et al., 2012a. Which is the best laparoscopic approach for inguinal hernia repair: TEP or TAPP? A systematic review of the literature with a network meta-analysis. Surg. Endosc. 26, 3355–3366.

Bracale, U., Rovani, M., Bracale, M., Pignata, G., Corcione, F., Pecchia, L., 2012b. Totally laparoscopic gastrectomy for gastric cancer: meta-analysis of short-term outcomes. Minim. Invasive Ther. Allied Technol. 21, 150−160.

Bridges, J.F., Mohamed, A.F., Finnern, H.W., Woehl, A., Hauber, A.B., 2012. Patients' preferences for treatment outcomes for advanced non-small cell lung cancer: a conjoint analysis. Lung Cancer. 77, 224−231.

Castaldo, R., et al., 2015. Acute mental stress assessment via short term HRV analysis in healthy adults: a systematic review with meta-analysis. Biomed. Signal Process. Control. 18, 370−377.

Chapman, R.H., Stone, P.W., Sandberg, E.A., Bell, C., Neumann, P.J., 2000. A comprehensive league table of cost-utility ratios and a sub-table of "panel-worthy" studies. Med. Decis. Making. 20, 451−467.

Cosh, E., Girling, A., Lilford, R., McAteer, H., Young, T., 2007. Investing in new medical technologies: a decision framework. J. Commer. Biotechnol. 13, 263−271.

Craven, M.P., 2007. Routes and requirements for realizing pervasive medical devices. In: Bardram, J.E., Mihailidis, A., Wan, D. (Eds.), Pervasive Computing in Healthcare, 2007. CRC Press, Boca Raton, pp. 217−241, Chapter 9, ISBN 978-0-8493-3621-8.

Craven, M.P., Morgan, S.P., 2011. Early stage economic evaluation with a small medical device start-up company using a Markov model. J. Med. Devices. 5, 027516-1.

CRH-BME Curricula Reformation and Harmonisation in the field of Biomedical Engineering, 2008. Project Number: 144537-TEMPUS-2008-GR-JPCR (2008-4527).

de Bekker-Grob, E.W., Ryan, M., Gerard, K., 2012. Discrete choice experiments in health economics: a review of the literature. Health Econ. 21, 145−172.

Directive 93/42/EEC (medical devices in general); Directive 2007/47/EC (amending 93/42 and equates the medical software to a stand-alone device); Directive 98/79/EC (in vitro diagnostic medical devices); Directive 90/385/EEC (active implantable medical devices).

Dong, H., Buxton, M., 2006. Early assessment of the likely cost-effectiveness of a new technology: a Markov model with probabilistic sensitivity analysis of computer-assisted total knee replacement. Int. J. Technol. Assess. Health Care. 22, 191−202.

Furst, D.E., Schiff, M.H., Fleischmann, R.M., Strand, V., Birbara, C.A., Compagnone, D., et al., 2003. Adalimumab, a fully human anti tumor necrosis factor-alpha monoclonal antibody, and concomitant standard antirheumatic therapy for the treatment of rheumatoid arthritis: results of STAR (Safety Trial of Adalimumab in Rheumatoid Arthritis). J. Rheumatol. 30, 2563−2571.

Gallego, G., Bridges, J.F., Flynn, T., Blauvelt, B.M., 2011. Predicting the future impact of emerging technologies on hepatocellular carcinoma (Hcc): measuring stakeholders preferences with best-worst scaling. Value Health. 14, A176.

Guyatt, G.H., Sackett, D.L., Sinclair, J.C., Hayward, R., Cook, D.J., Cook, R.J., 1995. Users' guides to the medical literature. IX. A method for grading health care recommendations. Evidence-Based Medicine Working Group. JAMA. 274, 1800−1804.

Health technology assessment of medical devices. WHO Medical device technical series, 2011 [http://whqlibdoc.who.int/publications/2011/9789241501361_eng.pdf].

Heitman, E., 1998. Ethical issues in technology assessment. Conceptual categories and procedural considerations. Int. J. Technol. Assess. Health Care. 14, 544−566.

IFMBE, 2012. HTA division webpage. Available: [http://www.ifmbe.org/index.php?option=com_content&view=category&id=53:health-care-technology-assessment&Itemid=167&layout=default].

Ijzerman, M.J., Steuten, L.M., 2011. Early assessment of medical technologies to inform product development and market access: a review of methods and applications. Appl. Health Econ. Health Policy. 9, 331−347.

Ijzerman, M.J., van Til, J.A., Snoek, G.J., 2008. Comparison of two multi-criteria decision techniques for eliciting treatment preferences in people with neurological disorders. Patient. 1, 265−272.

Jarm, T., Miklavcic, D., Pallikarakis, N., Bliznakov, Z., Magjarevic, R., Lackovic, I., et al., 2012. In: Jobbágy, Á., Magjarevic, R. (Eds.), Proposal for Generic Biomedical Engineering Programs Based on European Experience 5th European Conference of the International Federation for Medical and Biological Engineering, vol. 37. Springer, Berlin Heidelberg, pp. 1418−1421.

Keystone, E.C., Kavanaugh, A.F., Sharp, J.T., Tannenbaum, H., Hua, Y., Teoh, L.S., et al., 2004. Radiographic, clinical, and functional outcomes of treatment with adalimumab (a human anti-tumor necrosis factor monoclonal antibody) in patients with active rheumatoid arthritis receiving concomitant methotrexate therapy: a randomized, placebo-controlled, 52-week trial. Arthritis. Rheum. 50, 1400−1411.

Kind, P., Dolan, P., Gudex, C., Williams, A., 1998. Variations in population health status: results from a United Kingdom national questionnaire survey. BMJ. 316, 736−741.

Liberatore, M.J., Nydick, R.L., 2008. The analytic hierarchy process in medical and health care decision making: a literature review. Eur. J. Oper. Res. 189, 194−207.

Lumley, T., 2002. Network meta-analysis for indirect treatment comparisons. Stat. Med. 21, 2313−2324.

Mulye, R., 1998. An empirical comparison of three variants of the AHP and two variants of conjoint analysis. J. Behav. Decis. Making. 11, 263−280.

Nord, E., 1999. Towards cost-value analysis in health care? Health Care Anal. 7, 167−175.

Oppe, M., Devlin, N.J., Szende, A., 2007. EQ-5D Value Sets: Inventory, Comparative Review and User Guide. Springer.

Pallikarakis, N., Bliznakov, Z., Miklavcic, D., Jarm, T., Magjarevic, R., Lackovicm, I. et al. 2011. Promoting harmonization of BME education in Europe: The CRH-BME Tempus project. In: Engineering in Medicine and Biology Society, EMBC, 2011 Annual International Conference of the IEEE, pp. 6522−6525.

Pecchia, L., Acampora, F., Acampora, S., Bracale, M., 2007. A Multi Scale Methodology for Technology Assessment. A case study on Spine Surgery. In: 11th Mediterranean Conference on Medical and Biological Engineering and Computing 2007, Springer Berlin, Heidelberg, Vol. 2, pp. 762−765.

Pecchia, L., Bath, P., Pendleton, N., Jackson, S., Clarke, C., Briggs, P. et al. 2011a. The use of analytic hierarchy process for the prioritization of factors affecting wellbeing in elderly. In: presented at the International Symposium on Analytic Hierarchy Process (ISAHP), Sorrento, Naples, Italy.

Pecchia, L., Bath, P.A., Pendleton, N., Bracale, M., 2010. Web-based system for assessing risk factors for falls in community-dwelling elderly people using the analytic hierarchy process. Int. J. Anal. Hierarchy Process. 2, http://www.ijahp.org/index.php/IJAHP/article/view/61.

Pecchia, L., Bath, P.A., Pendleton, N., Bracale, M., 2011b. Analytic Hierarchy Process (AHP) for examining healthcare professionals' assessments of risk factors. The relative importance of risk factors for falls in community-dwelling older people. Methods Inf. Med. 50, 435−444.

Pecchia, L., Bracale, U., Bracale, M., 2009a. Health technology assessment of home monitoring for the continuity of care of patient suffering from congestive heart failure. In: Dössel, O., Schlegel, W.C. (Eds.), World Congress on Medical Physics and Biomedical Engineering, September 7−12, 2009, Munich, Germany, vol. 25/12. Springer, Berlin Heidelberg, pp. 184−187.

Pecchia, L., Bracale, U., Melillo, P., Sansone, M., Bracale, M. 2009b. AHP for Health Technology Assessment. A case study: prioritizing care approaches for patients suffering from chronic heart failure. In: presented at the International Symposium on AHP (ISAHP), Pittsburgh, Pennsylvania, USA.

Pecchia, L., Martin, J.L., Ragozzino, A., Vanzanella, C., Scognamiglio, A., Mirarchi, L., et al., 2013. User needs elicitation via analytic hierarchy process (AHP). A case study on a Computed Tomography (CT) scanner. BMC Med. Inform. Decis. Mak. 13, 2.

Pecchia, L., Mirarchi, L., Doniacovo, R., Marsico, V., Bracale,, M., 2009c. Health technology assessment for a service contract: a new method for decisional tools. World Congress on Medical Physics and Biomedical Engineering. Vol 25 (Pt 12), 105−108.

Reiser, S.J., 1988. A perspective on ethical issues in technology assessment. Health Policy. 9, 297−300.

Saaty, T., 1982. How to structure and make choices in complex problems. Hum. Syst. Manage. 3, 255–261.

Saaty, T.L., 1977. A scaling method for priorities in hierarchical structures. J. Math. Psychol. 15, 8.

Saaty, T.L., Vargas, L.G., 2001. Models, Methods, Concepts & Applications of the Analytic Hierarchy Process. Kluwer Academic Publishers, Boston.

Sacchini, D., Virdis, A., Refolo, P., Pennacchini, M., de Paula, I.C., 2009. Health technology assessment (HTA): ethical aspects. Med. Health Care Philos. 12, 453–457.

Scholl, A., Manthey, L., Helm, R., Steiner, M., 2005. Solving multiattribute design problems with analytic hierarchy process and conjoint analysis: an empirical comparison. Eur. J. Oper. Res. 164, 760–777.

Sloane, E.B., Liberatore, M.J., Nydick, R.L., Luo, W.H., Chung, Q.B., 2003. Using the analytic hierarchy process as a clinical engineering tool to facilitate an iterative, multidisciplinary, microeconomic health technology assessment. Comput. Oper. Res. 30, 1447–1465.

Sonnenberg, F.A., Beck, J.R., 1993. Markov models in medical decision making: a practical guide. Med. Decis. Making. 13, 322–338.

Sutton, J., 2000. Methods for Meta-Analysis in Medical Research. J. Wiley, Chichester; New York.

Van de Putte, L.B., Atkins, C., Malaise, M., Sany, J., Russell, A.S., van Riel, P.L., et al., 2004. Efficacy and safety of adalimumab as monotherapy in patients with rheumatoid arthritis for whom previous disease modifying antirheumatic drug treatment has failed. Ann. Rheum Dis. 63, 508–516.

Van de Putte, L.B., Rau, R., Breedveld, F.C., Kalden, J.R., Malaise, M.G., van Riel, P.L., et al., 2003. Efficacy and safety of the fully human anti-tumour necrosis factor alpha monoclonal antibody adalimumab (D2E7) in DMARD refractory patients with rheumatoid arthritis: a 12 week, phase II study. Ann. Rheum. Dis. 62, 1168–1177.

Weinblatt, M.E., Keystone, E.C., Furst, D.E., Moreland, L.W., Weisman, M.H., Birbara, C.A., et al., 2003. Adalimumab, a fully human anti-tumor necrosis factor alpha monoclonal antibody, for the treatment of rheumatoid arthritis in patients taking concomitant methotrexate: the ARMADA trial. Arthritis. Rheum. 48, 35–45.

9

Integrated Risk and Quality Management in Hospital Systems

Roberto Miniati[1], Francesco Frosini[1], Fabrizio Dori[1,2]

[1]DEPARTMENT OF INFORMATION ENGINEERING, BIOMEDICAL LAB, VIA DI SANTA MARTA, FIRENZE, ITALY [2]REGIONAL HEALTH TECHNOLOGY DEPARTMENT, TUSCANY, ITALY

Introduction

For many years, studying the safety and reliability of services in various fields has become the most important aspect and the main effort of several research groups. The "entrepreneurial risk" is proportional to the complexity of the systems themselves, and the involved variables are numerous. Such a concept can be extended also to the healthcare systems: the healthcare organization, in fact, is a very complex system characterized by an elevated multiplicity of variables, a high level of organization and interaction among the various hierarchical levels, and a strong dependence that the system itself has on individuals, be they patient or health personnel.

Recently, due to ever-increasing healthcare costs, which are based on the balance of economic refunds proportional to the carried out clinical activity by individual hospitals, the essentially typical concepts of quality, such as efficiency and productivity, have become ever more important for individual organizations and are considered to be at high risk. For such reasons, the achievement of acceptable hospital safety and quality levels sees, in addition to the application of classical risk management methodologies, the necessity to also adopt models and methodologies coming from the world of quality. This creates an integrated management of the two approaches: a process of higher quality is a safer process.

If initially the Risk Management activity had been associated exclusively with the theory of Clinical Risk Management (CRM), throughout the years it has been refined and concentrated mainly on the need to govern the cultural vehicle that permits the systematic tackling of the risk and its causes: the analysis of the processes; adopting models coming from research (Health Technology Assessment), or adopting models coming from other industrial sectors, from the sector of reliability engineering to the statistic simulation of complex processes.

In the matter of the cultural roots of the quality model, these can be found in the Total Quality Management only to later arrive at the approaches Six Sigma and Lean Thinking. These methods are used for the management of the process oriented toward the client, were developed in the United States and then in Japan, respectively, and are the syntheses of a series of reflections on the quality of products/services and on the organizational conventions suitable for obtaining this quality.

An integrated system of Quality Risk Management in hospitals, other than defending against risks, adds value to the organization and its stakeholders supporting the objectives of the organization through the improvement of the decision process, from the planning to the economic sustainability, by making the most efficient use of resources through the protection and strengthening of the hospital image and the creating of a system that permits the carrying out and control of every future activity.

Methods

Error Analysis and Patient Safety

Given the diagnosis and cure context in which process analysis and risk management must advance, error management developing in the "clinical risk analysis" sphere is placed alongside the approaches already seen introducing a point of view in which, against a striving effort to optimize the process, the concept of quality is developed as a lack of clinical error. The techniques used are, therefore, suitable for the need to make the potential adverse event in clinical activities emerge.

As a consequence of the necessity to observe the human errors from a new point of view, it is possible to distinguish active failures, which determine immediate consequences, from latent errors, which instead remain "silent" in the system as long as a certain triggering event does not render them evident in all of their potentiality, provoking consequences more or less serious.

Of late, it has been revealed that also the errors of organizational origin, which belong to latent errors, have a decisive role even though not all latent errors produce an active failure, nor do all errors provoke damage. The safety of the patient comes from, therefore, the capability to plan and manage processes capable of, on the one hand, containing the effects of the errors that are verified (protection) and on the other hand reducing the probability that such errors take place (prevention). So, two analysis typologies can be taken advantage of: reactive-type analysis and prospective-type analysis.

Reactive-Type Analysis

The reactive analysis calls for a study in retrospect of the incidents and is aimed at pinpointing the causes that permitted their occurrence in order to introduce corrective procedural changes. For this reason, a reverse analysis is led following the occurrence of the adverse event, in order to produce a reconstruction that, from the active failures, pinpoints the risk factors at the workspace and produces a final result, which aims to define the organizational causes that have generated them. The approaches mainly used include: Incident Reporting, Trigger Analysis and Review, and Root Cause Analysis.

INCIDENT REPORTING

Incident reporting is a collection of anonymous forms to warn about adverse events: through the alert report of errors and "possible errors," a series of fundamental information can be collected to trace the path which has allowed for the occurrence of the adverse event. In order to have a comprehensive evaluation of the phenomenon, it is necessary to define report standard systems, which are fundamental to the collection of information on which analyses and recommendations are based, even though they are actually effective among the highly suspicious operators who may be afraid of blame or punishment for signaling an error.

TRIGGER ANALYSIS AND REVIEW

The research of proof (trigger) to pinpoint possible errors that occur during the hospital care and medical process calls for an analysis of the patient and health records, even through computer alert systems, to locate the indicators of possible errors, are marked with a positive index (index of suspected error) and whose evaluation is left to experts.

ROOT CAUSES ANALYSIS (RCA)

The Root Causes Analysis (RCA) starts with the observed errors within a system. The causes are researched through an inductive method that proceeds by means of questions that explore the reason for each action and all of its possible deviations. The RCA first focuses on the system and its processes and later on the performance of the personnel. Similar analyses of the causes have to establish the human factors directly associated with the incident, sentinel events, or adverse events, determine the latent factors associated with them, and identify the necessary changes to avoid a repetition of the event.

Prospective-Type Analysis

The prospective analysis, instead, aims to pinpoint and eliminate the criticality of the system before the incident occurs, that is, its function is to reduce the incident and its consequences preemptively, analyzing the process in all of its phases. The studies can be either quantitative or qualitative and go on to analyze the process in its phases, using the process analysis techniques that are already sensible in other settings, of which the most used are Integration Definition Language (IDEF), Cognitive Task Analysis, Human HAZOP, SHELL, and FMEA/FMECA. The FMEA/FMECA, Failure Mode Effect Analysis, and Failure Mode Effect and Criticality Analysis, respectively, together represent the most commonly used model.

FMEA/FMECA ANALYSIS

Created 40 years ago in the United States in missile fields, it has been used for over 30 years in industrial sectors, such as automotive, aviation, and nuclear, and has been recently proposed as an instrument for risk prevention for healthcare organizations by the Joint

Commission International. The Department of Veteran Affairs (DVA) in collaboration with the National Center for Patient Safety (NCPS) has introduced the abbreviation HFMEA (DeRosier et al., 2002), that is, Health Failure Modes and Effect Analysis, to indicate the applications of the FMEA/FMECA in the healthcare context. The Joint Commission provides information about the following applicable steps (JCAHO, 2001):

- Identify and give priority to high-risk processes
- Annually select at least one high-risk process
- Identify the potential failure modes
- Identify possible effects for each failure mode
- For the most critical effect, conduct an analysis of the origin of the causes
- Redesign the process to minimize the risk of that failure mode or protect the patient from its effects
- Experiment or apply the redesigned process
- Identify and apply efficacy measures
- Apply a strategy to maintain the efficacy of the redesigned process in time.

At a quantitative level, the independent parameters for the FMEA/FMECA are those reported in Figure 9.1:

1. O (Occurrence): probability of the event of a noncompliance;
2. S (Severity): gravity/criticality of the failure;
3. D (Detection): possibility to disclose an onset of a noncompliance.

The product of the three parameters, which only takes on values between 1 and 10, defines a quarter of the synthetic index known as the Risk Priority Number (RPN), which needs to be the lowest possible and fall under the defined acceptability limits for the specific context.

In the latest analysis, once the new process/technology is implemented, the monitoring of the efficacy of the corrective action has to be guaranteed, repeating the analysis in order to verify that new damaged modalities of unacceptable RPN have not been inserted involuntarily.

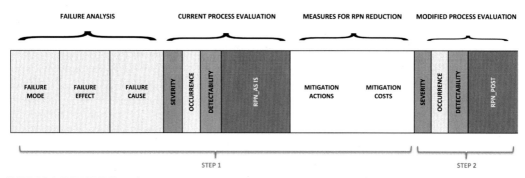

FIGURE 9.1 FMEA/FMECA tool.

Risk Management

The International Organization for Standardization (ISO) provides technical guidelines in many social and industrial sectors. The cases of the ISO 31000:2009 and ISO 31010:2009 concern principles and generic guidelines on risk management activities (ISO, 2009) and represent the base for every risk management application to any area of interest (ISO/IEC, 2009). Besides the classic risk management activity, seeing the high complexity of the system, the high criticality of the services provided in hospital combined with the ever-increasing technological dependence in some specified clinical areas, the concepts of Probabilistic Risk Assessment (PRA) and Business Continuity Management (BCM) are starting to be introduced.

Business Continuity Management (BCM)

The operative continuity is the combination of activity directed at minimizing the destructive effects, however damaging, of an event that has hit an organization or part of it, guaranteeing the continuity of the activity in general (British Standard BS 25999-2, 2007). The operative continuity consists of both the strictly organizational, logistical, and communicative aspects, which allow for the progression of an organizations' functionality, and the technological continuity that concerns the possibility to recover the service (Disaster and/or Business Recovery) (Hilles, 2007).

As reported in Figure 9.2, in a very similar manner to every risk management approach, the main steps for proper hospital BCM are:

1. Know the capacity and limitations of the services by way of an analysis of the clinical process;
2. Achieve an evaluation of the risk directed at the mitigation of the business vulnerabilities;

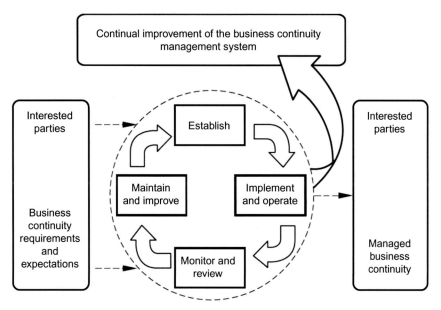

FIGURE 9.2 BCM framework.

3. Develop plans and projects to improve resilience;

4. Develop continuity plans, implement training and exercise activities;

5. Continuous updating of the plan (at least annually).

Probabilistic Risk Assessment (PRA)

Many studies using typical PRA methodologies have been successfully performed in hospital due to the high degree of complexity of the hospital system (Miniati et al., 2014), where individual functional areas are themselves interconnected in a multidimensional way and it is difficult to understand how one event can spread to the other areas. The traditional methods used are Boolean or probabilistic logic methods such as Event Tree Analysis (ETA) or deductive methods such as Fault Tree Analysis (FTA).

EVENT TREE ANALYSIS (ETA)

The ETA (Event Tree Analysis) methodology is an inductive process that starts at the TOP initiator event and defines the possible consequences with the aim to reduce the losses rather than prevent the event. As described in Figure 9.3, starting from a potentially dangerous condition of the system or from human error, possible developments in divergent forms are constructed, by inductive method, assigning a determined probability of the event (success "S" or Failure "F") to each choice and evaluating the final outcomes derived from the specific probabilities of success or failure of the various systems (e.g., "outcome A" = DEATH, "outcome B" = SEVEREINJURY,..., "outcome H" = NO CONSEQUENCES).

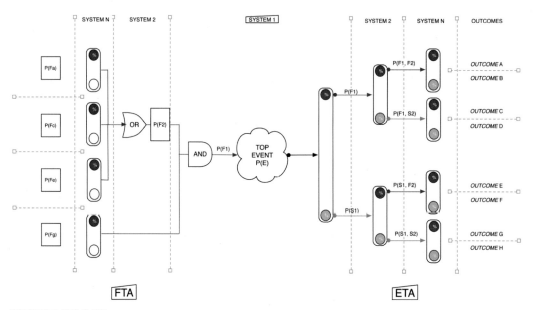

FIGURE 9.3 ETA & FTA.

FAULT TREE ANALYSIS (FTA)

The Fault Tree Analysis (FTA) is a complimentary technique to ETA, a top-down deductive failure analysis in which the reliability of a system (Top Event) is assessed by using Boolean logic to combine a series of failure probabilities belonging to the system's simplest events (the fault's leaves) (Verma et al., 2010). Safety and reliability aspects are the main application fields of the Fault Tree Analysis and can be used for designing safe systems (by identifying potential causes of failure) and for systems' breakdown prediction and diagnosis. Moreover, the FTA allows for the investigation of the an initial event's consequences on the top event of the analyzed system. In other words, by the failure probability of the basic elements, the top event probability of failure is estimated by taking into consideration the distribution of elements as well. Hence, FTA is used to calculate the overall probability of failure of a system with the reduction of the system into a group of both serial and redundant sections. The presence of parallel elements makes the system more reliable and the probability of failure decreases (AND Boolean gate), while on the other hand the presence of serial elements' distribution in the general system makes the top event probability of failure increase (OR Boolean gate), see Figure 9.3.

Health Technology Assessment (HTA)

The Health Technology Assessment (HTA) is an instrument that is more and more used and important within hospital structures. HTA is a systemic and multidisciplinary process that can represent a potential instrument of methodological support for the connection between "scientific" and "decision" settings (Battista and Hodge, 1999). Although at its inception the methodology focused essentially on the clinical context integrating Evidence-Based Medicine (EBM) and other traditional epidemiological methods, over the course of the years HTA has become ever more open to other healthcare applications, also including, in technical terms, organizational processes.

In fact, although the HTA started as a centralized activity to support political and national administrative decisions, in recent years, because of changes having taken place in several healthcare systems at a national level, the activity of HTA has become decentralized and the localized applications have risen at the level of individual hospitals defined as Hospital Based-HTA (HB-HTA). Such a process supports the hospital management that needs to decide based on economic-budgetary objectives of short- and medium-length periods, in addition to safety targets, efficacy, and organizational efficiency linked with consumer and employee satisfaction.

There currently exist several national and international associations with the objective to carry out studies and create collaboration networks between decision makers and all the other levels. Among these, the European EUnetHTA network has indicated 10 aspects to analyze during the carrying out of a proper evaluation (EUnetHTA, 2008). Among these, the "safety" deriving from the introduction of a specific technology (drugs, equipment, process) proves to be fundamental to pinpointing the possible impacts in the company context and the eventual possibilities of reducing risks introduced by such technology.

Total Quality Management, Six Sigma, and Lean Manufacturing

The evolution of the quality concept within healthcare structures has seen the development and use of various approaches that have followed the following methodological developments (Miniati et al., 2014):

1. Quality Assurance—Identify and eliminate defects;
2. Quality Control—Monitor and review the quality;
3. Quality Improvement—Minimize errors;
4. Total Quality Management (TQM)—Permanent quality improvement and technical quality;
5. Six Sigma—error-free quality.

The instruments mainly used in the healthcare setting belonging to both the TQM (with the exception of the FMECA already present in the previous sections), the Six Sigma, and Lean Manufacturing are reported subsequently.

Cause and Effect Diagram or Fishbone Chart and Pareto Diagram

In 1943, Kuaouru Ishikawa gave life to a diagram, which is now part of the cultural bases of an analytical and systematic approach to quality management. In the said diagram, given an effect, it is possible to identify the causes (cause-and-effect diagram or fishbone chart) through the identification of the studied effect and pinpointing and classifying the connected main and secondary causes, see Figure 9.4.

The Pareto diagram, named after its inventor, the economist Vilfredo Pareto, is the instrument to use when there are a lot of causes of which the relative importance to the system under consideration is the desired evaluation. The method tells us what the most recurring causes responsible for a certain effect are. Starting from an axiom *"few causes are mainly responsible for a given effect"* and it is therefore possible to concentrate all of the available resources only on these elements, disregarding the others.

SWOT Analysis

SWOT is an analysis that responds to the need to rationalize the decision processes as support to the definition of company strategies in contexts characterized by uncertainty and strong competitiveness. Moreover, it gives a visual and rational representation of the influence of some agents of exogenous nature (community politicians, economists) and endogenous (competition, power of the client and the supplier). It is based on a matrix divided into four fields, dedicated to the **S**trengths and **W**eaknesses appropriate to the context of the analysis and to the **O**pportunities and the **T**hreats of the external environmental context to which the organization is exposed.

Instruments Belonging to Six Sigma

The following methodology was introduced by Motorola in the 1980s and it is named after the statistical measure of the maximum variation (standard deviation of a normal

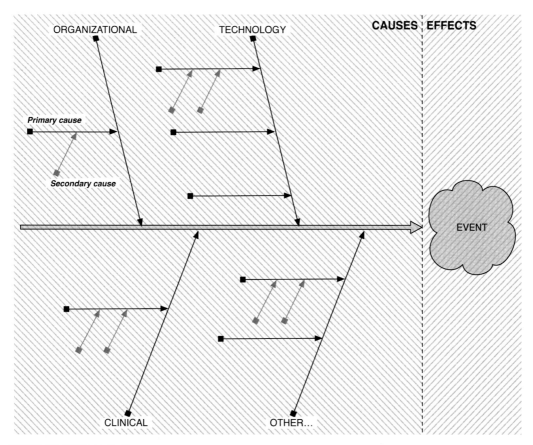

FIGURE 9.4 Fishbone chart.

distribution) of tolerance limits for defective products. Six Sigma can be applied to every health process providing measurable indicators for performance and quality assessment, from customer/patient expectations of reduced service costs to the compliance of the provided service with the current state of the art. The applied method to improve already existing and functioning processes is carried out according to the following steps (DMAIC): **D**efine the process/product to improve, **M**easure the current level performance of the process/product statistically, **A**nalyze the variability of the process/product, **I**mprove in relation to the objective of the product, and **C**ontrol the quality of the new process and support its maintenance. This methodology, suitable for any organization, is also based on the application of instruments already seen and characteristics of the TQM.

Lean Manufacturing (or Lean Production)

Identifying defects and wastes is absolutely not simple and entails a complete revision of the process establishing the satisfaction of the client receiving the product as the objective.

In particular, a unique line a thought has recently become used that goes very well with the Six Sigma philosophy: the Lean Manufacturing (or Lean Production) approach. Lean Manufacturing (LM) examines the process from the client's point of view (pull demand) dividing all of the operations that produce values from those that, on the other hand, do not add any.

Even though the healthcare applications present a difficulty to standardize the variability of a treatment path in productive lines, as requested by the productions typically Just In Time (Callahan, 2008), there are numerous positive examples in its hospital sector application (Tolkki and Parvinen, 2005). LM calls for a series of studies to be carried out in order to pinpoint the operations that create value and the operations without added value (Tucker et al., 1999) for the quality of the final product.

A fundamental indicator of LM is the Takt Time (TT), which defines the "beat of production." This represents the lapse of time that exists between the production of two consecutive units to satisfy the daily demand (Dc) considering a standard workday (H*S), see Equation (9.1).

$$TT = H * S/Dc \qquad\qquad (9.1)$$

where: H = available work hours/shift, S = number of shifts, and Dc = demand at capacity.

The TT is an important factor for the shaping of the production system and, together with the Cycle Time (CT), it allows for the definition of the number of necessary work cells to satisfy the demand in requested time (KnowledgeHills, 2011). The CT is defined as the work time necessary to complete the phase of the process, that is, how long the operative actually takes to carry out the transformation requested by the phase of work (White, 2011).

MUDA (THE SEVEN TYPES OF WASTE)

The main objective of the LM is, as mentioned earlier, to identify and eliminate the wastes. But, what is a waste really? How is it identified? Taiichi Ohno, the largest evaluator of the LM as well as the Executive Chief of Toyota, has defined the seven main types of wastes (MUDA in Japanese), which should be researched within a process. Later, an eighth was added: Defects, Overproduction, Transport, Waiting, Inventory, Motion, Overprocessing, and Skills. After correctly pinpointing the eight wastes, it is necessary to understand the process and analyze it together with the people who work it.

VALUE STREAM MAP

The construction of the Value Stream Map (VSM) allows for the identification of the material production's origin (Supplier), the final client (Customer), and the transformation activities (Activities) requested to reach the final form delivered to the client (Obamiro, 2010). The map uses symbols, terminology, and a scheme that permits seeing the process in its totality with the purpose of quickly pinpointing the evident wastes and generate improvements (from the Japanese term KAIZEN) in addition to evaluating the dynamics of information flow.

SPAGHETTI MAP

The Spaghetti Map (Morgan et al., 2012) analyzes the changes of the product/service during various production phases and it permits the pinpointing of personnel cluster points, overlap of workflow, and lack of coherence in the management of operations for the realization of the final product. To create a Spaghetti Map, the precise knowledge of both the workflow (so, the logical and functional process) and the knowledge of the spaces are necessary, and, therefore, the expert advice of specific operators is suggested.

Discreet Events Simulation Systems

In recent years, methodologies, deriving from the manufacturing sector, are being presented in healthcare, which are able to support the designer during the realization of efficient ways and shaped to the actual necessities. For this purpose, the discreet events systems have been revealed as suitable for the simulation of complex systems, such as healthcare itineraries (Waressara et al., 2013) through the use of statistic models like the Markov chain, queuing theory, or specific simulation software (Miniati et al., 2013).

Case study

Lean Application to Hospital Process

In this paragraph, the application of the Lean methodology is exposed to a generic service of pathological anatomy in hospital demonstrating the modality of application of the previously described instruments. The choice of the Pathological Anatomy service depends on the fact that it contains a high number of sub-processes right inside, has complex procedures (Frosini et al., 2014), and the necessity to have certain answers in brief periods, in addition to presenting an elevated risk of non-repeatability of the test in case of error (Burns, 2012).

In this case, the interest is represented by the biological sample extracted from the patient during the surgical activity, the client is the physician who needs to use the result of the biopsy to decide the therapy and the producer is the department of pathological anatomy.

The application of the Lean method, with the particular use of a Spaghetti Map and a VSM, has underlined some critical elements in the process such as useless movement of materials and personnel, in addition to showing the numerous operations whose sequence is interrupted by changing floors and crossing long hallways, see Table 9.1. Also within every room, numerous overlaps in workflow among various work cells are pinpointed, see Figure 9.5.

The construction of the VSM of the current process (AS IS) has called for the identification of all the main phases of the process estimating the Takt Time (TT), the Cycle Time (CT), and the necessary resources for each one. If there is no software that records the various phase times, the best solution for the order cycle estimate is to measure the performance directly on the field attempting to taint the actions and movements of the operators in the slightest

Table 9.1   Walked Distance and Storey Changes per Phase

Paths (Phase to Phase)	Distance (m)	Storey Change
Enter→Check-in	32	0
Check-in→Grossing	3	0
Grossing→Processing	50	0→ − 1
Processing→Embedding	35	− 1
Embedding→Cutting	75.5	− 1→0
Cutting→Staining	3	0
Staining→Reading	55	0→1
Reading→Results reporting	28	1→0

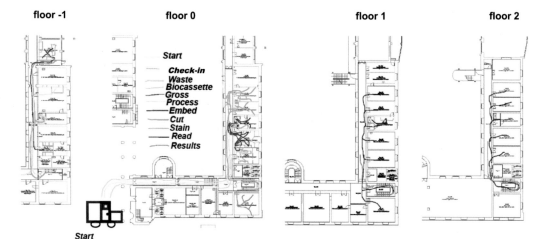

FIGURE 9.5 Spaghetti map.

way possible. As reported in Figure 9.6, the various phases have proven to be highly unbalanced among themselves and also with respect to the average value, creating a risk situation for the level of production quality. Considering that the TT is connected to the number of available resources (operators or technology composing the work cell/s within each workstation), a possible solution to its counterbalance could be the redistribution of the workforce among the various phases paying much attention to the typology of the personnel for the proper interchangeability among operators (medical, technical, administrative, or nursing personnel).

The application of the MUDA method (seven types of waste) has been carried out in synergy with the operators in every phase of work. In Table 9.2, the reported example is the form used for "Overproduction," intended as the modeling of the bottleneck, insofar from a leveled viewpoint it is advantageous to "produce only the necessary."

The application of the LM has produced interesting ideas for the planning of the future process (TO BE), such as the necessity to have a more rational integration of the production phases that permit the workflow of materials and personnel to continue in the most

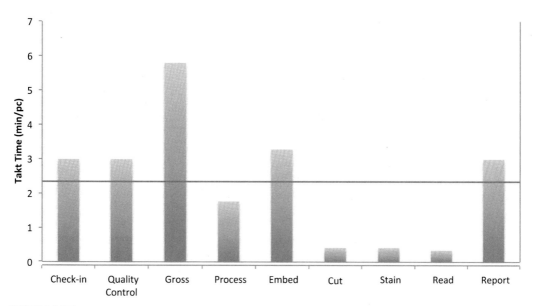

FIGURE 9.6 TT per process.

Table 9.2   MUDA Analysis—Overproduction

Phase	Level of Waste
Check-in	Low
Grossing	High
Processing	Medium
Embedding	Null
Cutting	Null
Staining	Null
Reading	Null
Writing	Null
Signature and Reporting	Null

optimized way, also including the arrangement of the various work cells. Furthermore, the objective to level the TT has established the right moment for a synergetic action between more professionals and hospital departments in addition to sharing technological investment plans and redistribution of personnel.

As a future development, considering that the TT is directly proportional to the order cycle, one of its variations can be obtained by modifying both the Cycle Time "CT" and the production volume. Considering that the volume depends on an external factor, such as the client's request, the CT variable is the only one that remains, whose eventual reduction, for example, can be obtained through a rationalization of the workstations' performance, through a 5S analysis and/ or operating the training in a way to standardize the schedules independent of the operators.

References

Battista, R.N., Hodge, M.J., 1999. The evolving paradigm of health technology assessment: reflections for the millennium. CMAJ. 160 (1010), 1464–1467.

British Standard BS 25999-2. 2007. Business continuity management. Part 2: Specification.

Burns, L.R. (Ed.), 2012. The Business of Healthcare Innovation, second ed. Cambridge University Press, Cambridge. Cambridge Books Online. Web. 04 September 2015. http://dx.doi.org/10.1017/CBO9781139176620.

Callahan, D., 2008. Health care costs and medical technology. In: Crowley, M. (Ed.), From Birth to Death and Bench to Clinic: The Hastings Center Bioethics Briefing Book for Journalists, Policymakers, and Campaigns. The Hastings Center, Garrison, NY, pp. 79–82.

De Rosier, J., Stalhandske, E., Bagian, J.P., Nudell, T., 2002. Using health care failure mode and effect analysis: the VA national center for patient safety's prospective risk analysis system. Jt. Comm. J. Qual. Improv. 28 (5), 248–267.

European Network for Health Technology Assessment, 2008. Work Package 4. HTA Core Model for Medical and Surgical Interventions v 1.0r.

Frosini, F., Miniati, R., Cecconi, G., Dori, F., Iadanza, E., Vezzosi, S. et al. 2014. Lean Thinking in Hospital: Case Study at the Pathology Laboratory. In: Proceedings from the 6th European Conference of the International Federation for Medical and Biological Engineering IFMBE. Croatia, pp. 613–616.

Hilles, A., 2007. The Definitive Handbook of Business Continuity Management. Second ed. John Wiley & Sons, Hoboken, New Jersey.

International Organization for Standardization 31000. 2009. Principles and Guidelines on Implementation.

International Organization for Standardization, International Electrotechnical Commission ISO/IEC 31010, 2009. Risk Management - Risk Assessment Techniques.

Joint Commission on Accreditation of Healthcare Organizations, 2001. 2002 Comprehensive Accreditation Manual for Hospitals: The Official Handbook (CAMH). Terrace, Oakbrook, IL.

KnowledgeHills, 2011. < http://sixsigma.knowledgehills.com/Operational-Cycle-Time-or-Takt-Time/a7p8 > (accessed 6.02.15).

Miniati, R., Capone, P., Hosser, D., 2014. Decision support system for rapid seismic risk mitigation of hospital systems. Comparison between models and countries. Int. J. Disaster Risk Reduct. 9, 1225.

Miniati, R., Cecconi, G., Dori, F., Frosini, F., Iadanza, E., Biffi Gentili, G. et al. 2013. A queueing theory based model for business continuity in hospitals. In: Conference proceedings from the Annual International Conference of the IEEE Engineering in Medicine and Biology Society, pp. 922–925.

Miniati, R., Frosini, F., Cecconi, G., Dori, F., Iadanza, E., Vezzosi, S. et al. 2014. Experience of Lean Six Sigma Quality Approach to Hospital Laboratory Services. In: Proceedings from the 6th European Conference of the International Federation for Medical and Biological Engineering IFMBE. Croatia, pp. 609–612.

Morgan, E.L., Nye, T., Bowen, J.M., Hurley, J., Goeree, R., Tarride, J.E., 2012. Mathematical modeling: the case of emergency department waiting times. Int. J. Technol. Assess. Health Care. 28 (2), 93–109.

Obamiro, J.K. 2010. Queuing Theory and Patient Satisfaction: An Overview of Terminology and Application in Ante-Natal Care Unit, Petroleum-Gas University of Ploiesti.

Tolkki, O., Parvinen, P. 2005. Primary and secondary technologies in radiology: a lean management perspective. In: International Conference on the Management of Healthcare & Medical Technology, 25–26 August 2005, Aalborg, Denmark.

Tucker, J.B., Barone, J.E., Cecere, J., Blabey, R.G., Rha, C.K., 1999. Using queueing theory to determine operating room staffing needs. J. Trauma. 46 (1), 71–79.

Verma, A.K., Ajit, S., Karanki, D.R. 2010. Reliability and Safety Engineering, 1st ed. Springer Series in Reliability Engineering. ISBN:978-1-84996-231-5.

Waressara, W., Pichitlamken, J., Subsombat, P., 2013. A generic discrete-event simulation model for outpatient clinics in a large public hospital. J. Healthc. Eng. 4 (2), 285–305.

10

Management of New Technologies: Software and Integrated Systems

Fabrizio Dori[1,2], Francesco Frosini[1], Roberto Miniati[1]

[1]DEPARTMENT OF INFORMATION ENGINEERING, BIOMEDICAL LAB, VIA DI SANTA MARTA, FIRENZE, ITALY [2]REGIONAL HEALTH TECHNOLOGY DEPARTMENT, TUSCANY, ITALY

Introduction

General concepts for Health Technology Management: From the Technical Rule to the Operative Context

The definition of "new technology" often runs the risk of going off into the banal, just to later bring about an influential bundle of meanings and situations rich with originality. Linked to this description, sure enough there are concepts of innovation, both technological and organizational, that make the necessity to still properly clarify what is truly "new" in technology the object of analysis.

Therefore, it is natural to face matters of innovation arising from various incentives and from which other concepts requiring in-depth analyses are branched out. The evolutionary panorama of the paths of diagnoses and care continually creates matters of innovation, and therefore of the "new," especially in the sense of the instruments present on the market. We can pinpoint the recipients of such instruments and pinpoint, for everyone, what research of the "new" implies:

- ... for the research:
  - a clear analysis of the needs;
  - a deepened knowledge of the market;
  - the adoption of the typical scientific methods of R&D.
- ... for the user:
  - new "instruments";
  - utilized for new procedures;
  - (often) within a new facility.

Clinical Engineering. DOI: http://dx.doi.org/10.1016/B978-0-12-803767-6.00010-6

- ... and for the manager (decision-maker), who is concerned with the task of handling:
  - the new management model (often as a consequence of the possibilities that new technologies offer);
  - the best evaluation capacity that can be arranged!

This panorama pushes us to face that which today is mainly "new" in technological terms.

The Innovative Systems

The increment of the technological progress, in particular the ever more frequent use of the IT systems in the field of clinical data analysis and of patient treatment control, has brought about the contemporary use of more EM devices and/or systems with the prevalent purpose of increasing the features. Among the examined concepts, there will be reference to the systems in general, focusing on the complex systems and the Programmable Electrical Medical Systems (PEMS).

The PEMS—Programmable Electrical Medical Systems

When the manufacturer does not conceive the devices to work together, it is necessary that someone assume responsibility for all the devices to work satisfactorily in the integrated system. This is a basic concept that lies in the widest definition of "state of the art," that is, all techniques considered correct for the "safe" implementation of a production.

The integration of systems can bring about the introduction of new dangers caused by the added operational connections and by those linked to the management software; it is fundamental that the person who assumes responsibility for the integrated system project supervise in a way befitting these dangers.

Using the legislative and normative framework of the EU context as an exemplifying and non-exhaustive claim and focusing, for the above reasons, on the case of the PEMS, we mention that putting a medical device in practice, and therefore also a PEMS, implies the product has to conform to the essential security requirements defined by the European Directives (Council Directive 93/42/EEC, 1993 and Council Directive 2007/47/EEC, 2007) in a way that does not cause danger to the operator, patient, or present personnel. A fundamental element is the compliance evaluation of a PEMS in order to assure the correspondence with the dictated requirements from the above-mentioned directive and permit its inclusion on the market.

For this purpose, the necessities of the CE branding for software and systems and, in particular, for the software for medical use contained in the PEMS, were studied. Once this necessity was confirmed, the method of *presumption of compliance* was chosen, which implies the identification of the reference Directive, of the legislative context and the normative one applicable to the system.

It is through the correspondence of the provisions with the various harmonized, technical Norms that come to create a state-of-the-art product that presumably conforms to the essential safety requirements.

The most important aspect for the planning purposes of a safe device/system is the capability to manage, control, and evaluate the risks, both present and potential ones; in particular, it is conceivable that for the correct risk evaluation within the PEMS, it is necessary to consider both the functional ones, data from the connection of various appliances, and those linked to the natural system software.

Analysis of the Risks Due to the Software

As it was outlined, the PEMS contain an important component software that permits the management of the system and of all the involved devices; it is the software that entails the greatest criticality for the planning purposes of the whole system. For this reason, it is important to evaluate all of the risks, in particular those due to the software component, scarcely considered in the past.

This is why the software belonging to a medical device is itself considered a medical device and, therefore, needs to conform to the European Directives, and this entails an evaluation process of the risks linked to the software aspects that prove rather difficult.

This analysis follows a logical course, which starts at the functional nucleus of the system: at the management of the risk derived from the PEMS software, which needs to be accompanied by another activity dedicated to the management of the risks linked to the functional connections of the PEMS. Considering this leads to the safety of the entire system (Figure 10.1).

The conformity regarding the software safety is verified by applying the provisions of the general norm EN IEC 60601-1 (IEC, 2012) on fundamental safety and on essential provisions, and putting the risk management process described in the norm EN ISO 14971 (ISO, 2009) in place.

Therefore, the project of a PEMS needs to be developed within the risk management process, in which the risk control measures relate to the risks in need of check.

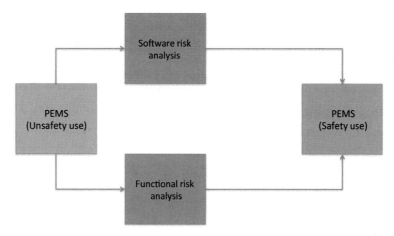

FIGURE 10.1 Analysis of the risks of the PEMS.

Furthermore, in order to be able to thoroughly develop the PEMS management software in compliance with the essential safety requirements, it is necessary to complement the norms mentioned above as well as the norm EN IEC 62304 (IEC, 2006) related to the life cycle of the software for medical devices.

The described processes, activities, and tasks form a common structure for the whole life cycle of the considered software.

Maintenance Procedure

Through the integration of these international standards, it is possible to develop a process for the correct development and maintenance of the software for medical use that aims to manage the software product in all of its life cycle phases.

In addition to permitting the planning of a software product in compliance with essential safety requirements, this procedure proposes to also produce documentation of risk management, which, upon being studied, reveals the evaluation of the conformity.

The process of developing and maintaining the software for medical use proposes to:

- be objective;
- be simple to apply; and
- be versatile.

Necessity of the CE Branding of the Software and of the PEMS

With the synthetic description of the systems, with and without a programmable component, it has been possible to extract some fundamental aspects that determine the importance of the CE branding for the EM systems and particularly for the PEMS, in which there is a relevant, additional aspect present, that is, the presence of an important software component.

Complexity of a System

It has been said that for inserting a product on the market, it is necessary that it respects the essential safety requirements and this is extremely easy for a simple device.

An orthopedic insole is classified by the pertinent Directive as a "fitted medical device," it does not require the application of the CE branding, but it still needs to be realized abiding by the essential safety criteria; and in this case, it is sufficient to assure that the material(s) used for manufacturing be compatible with the individual, that is, they present characteristics of being nontoxic, nonallergenic, biocompatible, and reliable to the request.

The EM systems in general, in order to increase the features, present a larger complexity and so it will be more complicated to successfully evaluate the consideration of the essential safety requirements inasmuch as it is legitimate that there will be more types of associated dangers.

Appendix D of the international standard EN ISO 14971 (ISO, 2009) reports an incomplete list of dangers associated with a medical device/system divided into:

- dangers linked to energy;
- biological dangers;

- environmental dangers;
- dangers related to an incorrect outlet of energy and substances;
- dangers related to the use of the system;
- dangers derived from an inappropriate, inadequate, or excessively complicated user interface;
- dangers derived from malfunction and failure, maintenance, and aging.

It is therefore predictable that to assure the conformity with the essential safety requirements, it is preferable to follow the applicable indications and regulations provided by norm technicians for the achievement of a "state-of-the-art" system.

Importance of the Software

Another aspect to consider is that the technological progress tends to transfer as many features as possible toward the device software for two reasons:

1. the first reason is ascribable to purely economical aspects and the aim is the elimination of some "elements" whose features are reproduced by the software;
2. the second is for the versatility of the system, that is, by delegating the intelligence of the entire system to the software, it becomes very easy for the manufacturer to restore the entire system through the addition of features in the chief software.

The software, therefore, becomes an innovative aspect of any device and system, but it is necessary to highlight that the centralization of the features consequently brings about the increase in its criticality, inasmuch as carrying out a complete evaluation of the risk while considering the dangers linked to software aspects, and therefore, evaluating the conformity with the essential safety requirements becomes ever more complicated.

Precisely due to the difficulty in carrying out a complete evaluation of the risks, the decision features of the software are often intended as support (and advise) features for the operator, who, for heightened safety, has to manually validate the final decision, inasmuch as an error in these decisions can bring about extremely dangerous and uncontrollable events.

To discuss this, it is possible to think of a software that has the feature of setting some parameters of a vital device. In order to minimize the risk that the software sets erroneous or damaging parameters, it is possible to introduce a limitation of parameter types or to require the confirmation of the operator.

Different Methodological Approaches

A further aspect that confirms the necessity of the CE branding of a software that works as a medical device is expressed in the different methodological approaches existing among those that deal with medical devices and software planning.

Obviously, these professional figures will have various competencies, but it is necessary in any case that there be an addressing on how the programmer should operate to make sure the risk management aspects be controlled during development, inasmuch as the addition of safety is troublesome after completing the development.

Methods

The software development process has to be initially defined and show a methodology that permits safely carrying out the activities. Inside, a series of milestones for each of the activities has to be defined, including those of risk management, which have to be completed before passing to the following milestone. The necessity to establish the steps to be fulfilled assures that the needed consideration has been given to all of the activities, in particular, the things that are necessary to do before starting the development and the outcome that the development shows. Also in the planning phases of the maintenance procedure, it is important to define what to do, why, and how to proceed.

What to do:

1. Define the critical points of the process
2. Evaluate the associated risk at every critical phase
3. Pinpoint possible solutions to single criticalities
4. Review the entire process after the "solutions"!

Why:

1. Organize ("use common sense");
2. Share the language and the process;
3. Make the criticalities and solutions objective and shared.

How:

1. Description of the process, listing phases, sub-phases, connections, and participants
2. Elements for the evaluation of the risks identifying the risks based on an evaluation criterion
3. Criterion of acceptability
4. The semi-quantitative elements and indicators
5. The methods of analyzing the processes

The Correct Software Development as a Base for PEMS Maintenance

During the development, and thereby the production, of an electromedical appliance or system in which a software component is present, it is necessary to be able to implement a correct and complete risk management without overlooking any aspect of the product, taking into consideration the other processes that allow the checking and management of additional features of the device.

The fundamental node for the production of PEMS is the correct development of the software for medical use, which is at the base of the device itself; in fact, it is through this subsystem software that the PEMS features are decided.

The Norm EN IEC 62304 (IEC, 2006) describes the processes related to the life cycle of the software for medical use that permit the development of a software product on which an accurate and exhaustive risk management was carried out.

These processes were described in the previous chapter and they are:

- Process of software development;
- Process of software maintenance;
- Process of risk management;
- Process of managing the software configurations;
- Process of resolving software problems.

As mentioned, the whole of these processes constitutes the life cycle of a software for medical use and during the product's development, there is no specific temporal order on which process proceeds first, but various phases of each will be run depending on the occurring problems, on the carried out modifications or on the variations of the software configurations.

The fundamental issue is that appropriate procedures that verify and control the development of the device, and particularly of the software for medical use, are present to guarantee the fundamental safety and the essential settings defined in the project phases.

To define this procedure, Article 5.1 of the international standard EN IEC 62304 (IEC, 2006) has been very useful, which refers to the definition of a software development plan.

Risk management is a fundamental process and is run at the end of every activity of the *Process of software development*, along with the *Process of resolving software problems* for the problems found during the development of the product.

To use the *Process of managing software configurations*, the software configurations and the managing of their modifications need to be defined, also including SOUP and support software, and furthermore, the following need to be considered:

- classes, types, categories, or lists of the items to be checked;
- activities and tasks for managing the software configuration;
- the facility in charge of carrying out the management of the software configurations;
- their realization with other facilities, as with software developers and maintenance technicians;
- when the items are placed under configurations checks;
- when the process of resolving problems is used.

The software development needs to be realized at the development of the system, and therefore the software requirements and the PEMS requirements need to be defined. The manufacturer needs to define the procedures to coordinate the software development, the project, and the development of the validation.

It is necessary to keep the traceability among the system requirements, software requirements, software systems tests, and the risk control measures implemented in the software.

This development plan always needs to be updated during the cycle of development of the software for medical use.

The classification of each software system and software item is also important in safety classes depending on the possible effects on the health and safety of the patients and/or operators, inasmuch as a software system in Class C requires a higher reliability with respect to a software system in Class A. In fact, for software items in Class C, the manufacturer needs to define standards, methods, and instruments that will be used.

These considerations are important for the identification of a process for the correct development of the software product and an analogous procedure for its maintenance, in which the *Process of software maintenance* will also be taken into consideration.

Procedure for Proper Maintenance

Once the software product, or, more generally, a PEMS, has been inserted into the market, there are necessary, appropriate procedures to verify the correct operation of this and in case of warnings of encountered anomalies, the manufacturer is compelled to evaluate such defects to assure that they do not compromise the health and safety of the patients and/or operators.

In the post-production procedure, there will be processes present that identify the occurring problems and fix them through proper maintenance and vice versa, a maintenance process that identifies potential problems before they come to be identified and/or generated by the product.

Discussion

Through these considerations, the way in which these processes interact among themselves has emerged, thereby arriving at the aim of the work, that is, a procedure for the correct development and maintenance of the software for medical use.

From the analysis of this procedure, some significant aspects are deduced:

1. The importance of risk control during the project phase inasmuch as a late risk evaluation leads to a burdensome commitment because of the implementation of possible additional safety measures.

   These entail a further risk evaluation in order to guarantee that their implementation within the PEMS does not create additional dangers; moreover, it is rather demanding to keep an opportune traceability of the dangers in the risk management report if these are evaluated only as finished products.
2. The procedure makes use of the additional processes that support the software development in order to completely manage all of the risks that can be created.

   In the developing procedure, it is necessary to highlight that a problem encountered by the *Process of resolving software problems* is resolved by the *Process of risk management*, and that the results from this can lead to a modification of the software configurations which in turn can bring about additional risks.

   In the maintenance procedure on the other hand, it is the *Process of software maintenance*, together with the problem resolving one, that pinpoints problems and feedback revealed in the final product and "passes" them to the *Process of risk management* to analyze possible modifications that belong to the following version of the software product.

3. The development of the software is directly correlated to the development of the system, inasmuch as it is from PEMS regulations that the software requirements, which are the starting point for the definition of the software product, come to be defined.

    Let it be pinpointed that during the check processes, new risk control measures or additional safety features that can modify the regulations of the PEMS can be defined; it is for this reason that the system regulations are reviewed whenever the risk analysis finishes.

    It is noted that due to the elevated importance of the *Process of risk management*, the development of a software for medical use starts again only after the evaluation, when there is no need for an additional reduction of the risk.

4. During the development of the procedure, part of the documentation of risk management is produced, which is of fundamental importance for the evaluation of the conformity with the general standard, inasmuch as it has a traceability on the carried out processes, encountered dangers, and measures of risk control applied in order to reduce or eliminate the residual risk.

For a qualitative evaluation of the development procedure of a software for medical use, it is opportune to carry out a SWOT analysis, that is, pinpoint the strengths, the weaknesses, the opportunities, and the threats.

The aim of this procedure is to allow the development of a software for medical use with the appropriate documentation.

The strengths of this procedure can be summarized in three adjectives: simple, versatile, and objective.

Simple: This application does not require large mental strains because the procedure guides the user step-by-step on the carried out processes and activities for a proper development of the software.

Versatility: The procedure has not been studied for the overviewed problem, but has a general and polyvalent character and it is useable both for PEMS management software development and *stand-alone* software starting directly from the software requirements. It is sufficient to modify the regulations for the PEMS in question or the software requirements that have the adaptability for the "state-of-the-art" development of the most varied software, not only for medical use.

Objectivity: Referring particularly to the process of risk management, it is highly probable that from the system description and its expected use/objective, two different users of the procedure come to identify the very similar dangers inasmuch as this process is repeated various times during the planning.

The main disadvantage of this procedure is that the need to create a safe software product and therefore, the need to consider all of the possible dangers that can be run into requires an accurate risk control phase. This need is expressed in the application of the

Process of risk management in a cyclical and continuous way after every activity of the development process, involving an excessive amount of time by the end of the planning.

While the *Process of risk management* is structured in such a way as to produce similar results independent of who applies them, this guideline allows flexibility to the manufacturer on the definition of the milestones, on how to carry out the tests for the integration and on how to carry out the verification of the achieved software system.

This opportunity allows the manufacturer to evaluate the best way to operate depending on the problems presented in front of him, allowing for a wider flexibility of the procedure.

An evident obstacle of the procedure is that its destination, especially in the final phases of the breakdown process, is referred to personnel with the competency of a software developer.

The phases referring to the analysis and the risk evaluation are, instead, more suitable for those that have knowledge of risk management, or better yet, for someone who deals with the possible consequences of the occurrence of a danger.

It is precisely from the necessity of both pieces of knowledge that the application of this procedure brings about the comparison of two different methodologies on how to face the problems.

Conclusions

The current panorama in which technologies move presents two elements that inevitably have to be faced in the immediate future of facilities appointed to the 360° management of the diagnosis and cure services, a future nonetheless already "present" regarding the needs of the users, but still in improvement phases on the part of managers.

The system definition is variable and complex. In particular the large flexibility and versatility of such "technological systems" represent, on the one hand, one of the best characteristics for their diffusion and implementation; on the other hand, they are the cause of the difficulty in their management inasmuch as it is difficult to "freeze" a specific configuration.

Such complexity has, in any case, the necessity to be governed to avoid, on the one hand, the presence of potentially dangerous systems and, on the other hand, maybe this is the most critical element, the emergence and the reinforcement of the certainty that technology proceeds "too fast to be controlled" and, therefore, that a worrisome superficiality hides behind the screen of the "technical progress."

Also in this case, we believe that the answer cannot be a method capable of analyzing the ongoing processes and following the developments even with new approaches and management models that are ever more suitable to the real needs of the expected use of the technology: the cost is undeniably high because it also, and especially, regards the capacity of modifying our approach to innovation.

References

Council Directive 93/42/EEC. 1993. Concerning Medical Devices.

Council Directive 2007/47/EC. 2007. Amending Council Directive 90/385/EEC on the approximation of the laws of the Member States relating to active implantable medical devices, Council Directive 93/42/EEC concerning medical devices and Directive 98/8/EC concerning the placing of biocidal products on the market.

International standard IEC 60601-1:2005 + AMD1. 2012. 2012 Medical electrical equipment Part 1: General requirements for basic safety and essential performance.

International standard IEC 62304. 2006. Medical device software — Software life cycle processes.

International standard ISO 14971. 2009. Medical devices — Application of risk management to medical devices.

11

Clinical Engineering and Disaster Preparedness

Yadin David[1], Fred Hosea[2], Cari Borrás[3]

[1]BIOMEDICAL ENGINEERING CONSULTANTS, LLC; UNIVERSITY OF TEXAS SCHOOL
OF PUBLIC HEALTH, HOUSTON, TX, USA [2]KAISER PERMANENTE, CLINICAL TECHNOLOGY
(RET.), CONVIVIA, PRESIDENT [3]RADIOLOGICAL PHYSICS AND HEALTH SERVICES,
WASHINGTON DC, USA

Introduction

New Dimensions of Professional Practice for Clinical Engineers

Disaster management is one area among several where the biomedical and clinical engineering professions have significant opportunities to expand their roles in organizational and institutional preparedness, in such areas as research, community networking, program development, scenario development, pilot testing, early situation awareness, systems vulnerability assessment, recovery support, consultation, cooperation with other professionals inside and outside the hospital, and leadership. These activities can be significant additions to the day-to-day operational concerns that often define the profession. Biomedical and clinical engineering professions are often more historically informed and the most systemically experienced professionals to provide expertise in managing the lifecycle of medical devices, systems and related processes. They understand clinical workflow vulnerabilities and practical alternatives, they may manage supply-chain relations with vendors and manufacturers, and they have close working relations with physicians, nurses, and facility administrators. This breadth of experience can be a unique and valuable connective resource in disaster management plans and capabilities, especially in circumstances where preventing infrastructure damage requires planning and setting up alternate sites of care in schools, stadiums, and refugee camps.

Clinical Engineering for Disaster Management Is Embedded in Multiple Systems

Many institutions and professions around the world have devoted years to planning, assessing, preparing, responding, and recovering from disasters. However, many of these plans may be outdated, especially if they have failed to take advantage of new technologies.

Clinical Engineering. DOI: http://dx.doi.org/10.1016/B978-0-12-803767-6.00011-8

A disaster manual or procedure may be sitting on a shelf somewhere, but it may not be accessible when a disaster occurs (e.g., it may be buried under rubble), requiring people to improvise their activities in an *ad hoc* fashion that wastes precious time and resources. As the next generation of wireless communications and clinical technologies emerges, there will be many new opportunities to make disaster response more accessible, sophisticated, and reliable. These new capabilities also introduce new gray areas of professional practice and responsibility that will need to be evaluated and assigned to the most appropriate professional group. Clinical engineers should be closely involved, ready to share their expertise, define new roles, and expand their career paths.

For our purposes here, we recommend that clinical engineers have a global framework for understanding the "nesting" of disaster management capabilities, working downward from international institutions to national organizations, states, counties, cities, and local organizations.

Overview of Disaster Types Its Management and Terminology

Natural and man-made disasters have increased in their frequency and impact over the past 20 years (Leaning and Guha-Sapir, 2013) with annual average of reported natural disasters of 394 events per year with estimated annual economic losses from these disasters of US$ 143 billion (CRED, 2012) (Figure 11.1).

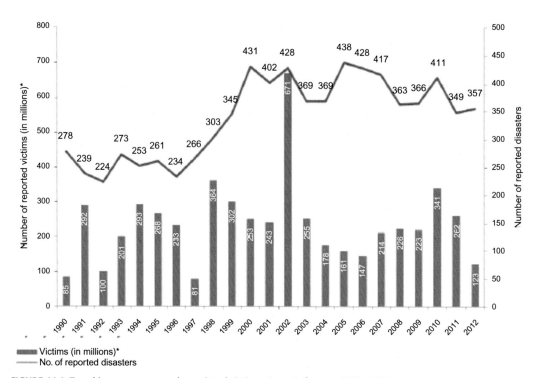

FIGURE 11.1 Trend in occurrences and associated victims. *From Reference CRED, 2012.*

With such a vast global impact, it is critically important to promote the awareness to the consequences of not being prepared for such events, to increase personnel training and preparedness on how to implement contingency plans for the staff and the infrastructure in facilities such as hospitals and regional clinics. Hospitals are expected to have a critical life-saving role, especially for events in the surrounding community. For that, staff need to be trained, facilities should be strengthened, and action plans have to be clearly implemented.

This chapter provides information on the variety of vulnerabilities faced by healthcare facilities exposed to both, natural phenomena such as earthquakes, flooding, and high-winds risks, and man-made occurrences such as radiological or nuclear incidents. It also describes the methodology to assess and mitigate the risks of damage and disruption of operations that can be caused by these events. The paper intends to be part of a training package aimed to expose healthcare facility personnel to disaster concepts, to prepare their facility and services to sustain functionality during such events, and to describe the methodology of hazard assessment prior to, during, and post the disaster event. The focus of the training is on addressing the needs of safely managing various technologies within the healthcare facility and on planning for patient and staff protection.

What Are the Disaster Stages?

Recent advances in monitoring and forecasting conditions leading to disaster have not yet shown their impact on reduction in community losses. While these efforts will continue to improve, they do not possess the ability to stop or delay the event. Therefore, understanding the basic disaster categories and the effect of each one of them will help hospital or clinic personnel to better focus their limited resources (Figure 11.2).

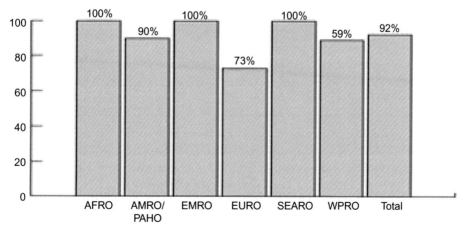

FIGURE 11.2 Countries reporting emergency disaster experiences in the past 10 years, by WHO region.

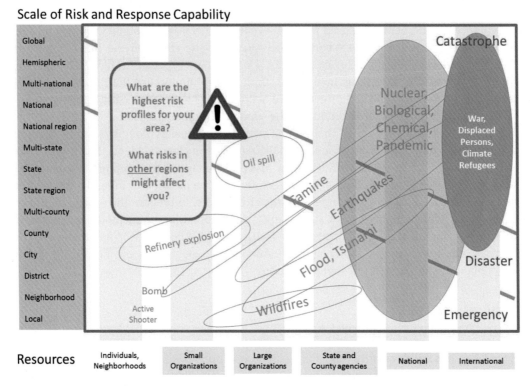

FIGURE 11.3 Scale of risks and necessary resources.

What is a disaster? A disaster is an unforeseen and sudden event that causes great damage, destruction, and human suffering. Disasters occur in many different forms with several levels of intensity and duration. Figure 11.3 demonstrates a gradation of event types, ranging from emergencies to disasters to catastrophes, which will all require different degrees of resources and preparations that need to be organized at local, state, national, and international levels.

A key companion concept for disaster management is "resiliency," defined in *Disaster Resilience: A National Imperative* (2012) (Disaster Resilience, 2012) as "the ability to prepare and plan for, absorb, recover from, and more successfully adapt to adverse events. Enhanced resilience allows better anticipation of disasters and better planning to reduce disaster losses—rather than waiting for an event to occur and paying for it afterward."

Resilience emphasizes a more proactive, "can-do" culture for individuals, families, and local jurisdictions that helps them maximize attitudes of local confidence and self-sufficiency as a foundation that will undergird the more formal, governmental, and other institutional efforts to manage disaster conditions. It is essential that preparedness at all levels helps people reduce citizens' unrealistic dependency or unmeetable expectations that higher level authorities will not be able to accommodate under many disaster circumstances.

Disaster preparedness is the process (Businessdictionary, 2014) of ensuring that an organization—together with its institutional partners—has reached a state of readiness to face effects of disastrous event and can continue to provide services by implementing proper preventive measures (David et al., 2013). This includes adequately trained personnel, practiced leadership, and clear planning for utilities and supplies that are essential to successfully sustain and protect operations during disasters. Disasters are primarily a health and social issue. In the last decade, more than 24 million people in Latin America and the Caribbean have lost their lives, loved ones, homes, workplaces, and possessions to natural or man-made disasters. Disasters have damaged or destroyed hospitals and healthcare facilities, leaving many people without access to health services critical to life. Classifications of natural disasters are: (i) Biological, (ii) Geophysical, (iii) Climate-related, and (iv) Meteorological. There are also classifications of man-made disasters, such as nuclear reactor failures, corrupt civil engineering and building construction practices, hydrological engineering miscalculations, preventable industrial accidents, and terror attacks. Disaster stages are: (i) preparation before the event occurs, (ii) impact during the event, and (iii) recovery after the event.

One of the biggest challenges of disaster and survival preparedness is convincing oneself, and funding authorities, that the risk does indeed exist for the possibility of disaster, and that making and investing resources in planning, preparations, and contingency plans are needed, and are a kind of life insurance.

Impact during the disaster is the stage at which the disaster strikes and the preparedness and contingency plans take effect. By now, hopefully, one has developed a hazard vulnerability assessment (HVA) and prioritization plan (ahead of time) containing estimates of the possible effects from the disaster on each one of the systems or equipment and how it is to be managed (David and Dreps, 2006). The impact stage will be frightful and shocking to the unprepared, and can result in the inability to take action, and/or will lead to poor decision making and uncoordinated efforts. Many affected people still simply sit and wait for "help." The vast majority of the public assumes and depends upon the government (or upon agencies within the system) to save them or help them, but in any hospital the first response by healthcare technology managers is critically important. During "impact," it is important to prioritize which of the medical technology and utility systems and equipment are to be addressed and secured. There is a need to recognize, understand, and assess what has happened, to estimate the consequences, and to gauge your responses and actions accordingly. In best cases, a hospital will have a formally constituted Emergency/Disaster Preparedness team with key representatives from the organization who meet over time, with expert consultations, to develop a preparedness and response plan, including members of the community and mutual aid agencies who will help each other to prepare for and respond to disasters.

Major advances have been made recently in software tools that can be used to capture and track critical information during the early moments and hours of a disaster. This "early situation awareness" can be of critical value in helping responders decide where and how to respond. The "Common Operating Picture" platform enables emergency/disaster organizations to collect real-time information on a shared, visual mapping tool that is based on zoomable Google Map technology that enables types of information to be displayed and be clicked on or off as needed. Local, regional, or national databases can be used to pre-populate the map with critical infrastructure information (gas lines, earthquake faults, flood zones, highways, hospitals, schools, evacuation routes, etc.). An open-source application can be downloaded onto smartphones, enabling trained citizens and first responders to send in real-time photos, audio commentary, classification by checklists, and assign special purpose icons to the event. Disaster managers at a central monitoring Command Center can click on icons as they pop up on the map and learn what conditions are being reported by exact GPS location, by neighborhood, region, and state.

After the disaster, the recovery stage begins when equipment and system status must be evaluated and preparations for the services to be patient-ready again should be taken. Documentation of the damage is a priority, including the use of photography for preparing repair/replacement plans. Common Operating Picture tools can be used to document damage assessment activities, using pre-loaded CAD drawings for a hospital, combined with verbal commentary and photos from smartphones. For longer-term recovery efforts, the Sahana Eden Humanitarian software platform (Sahana Software Foundation©) may be used to manage staff, volunteers, projects, replacement of damaged assets, track supply shipments, and query warehouse inventories. Participating organizations can pre-register their organizations to create regional databases of available hospital beds and other disaster resources that may be shared as appropriate.

Throughout these stages, the concern for staff well-being, rest periods, feeding, rotation, and needs of personal hygiene or medications must be addressed as well.

How to Prepare?

Every healthcare organization must have an emergency management program (Disaster Management Plan or other title) for the purpose of patient/resident/client care continuity in the event of emergency situations (Joint Commission, 2006). Healthcare organizations that offer emergency services or are designated as disaster receiving stations must have a program that addresses both external and internal disasters (Joint Commission, 2013). The program should be general and allow specific responses to the types of disasters likely to be encountered by the organization and applies to business occupancies. The program is based on the priorities identified in the hazardous vulnerability analysis (HVA) (Joint Commission, 2008). Based on an evaluation of incident probability/frequency specific to the organization, disasters that might be considered in an organization's plan include natural disasters, including the following types:

- Meteorological disasters: cyclones, typhoons, hurricanes, tornadoes, hailstorms, snowstorms, and droughts;

- Topological disasters: landslides, avalanches, mudflows, and floods;
- Disasters that originate underground: earthquakes, volcanic eruptions, and tsunamis (seismic sea waves);
- Biological disasters: communicable disease epidemics and insect swarms (locusts).

Man-made disasters, including the following types:

- Warfare: conventional warfare (bombardment, blockade and siege) and nonconventional warfare (nuclear, chemical and biological), population displacements from religious, ethnic, tribal, and insurgent violence;
- Civil disasters: riots and demonstrations, strikes, persons displaced by famine, drought, ecological hazards;
- Criminal/terrorist action: bomb threat/incident, nuclear, chemical, or biological attack, hostage incident, ongoing drug cartel hostilities and widespread gang violence;
- Accidents: transportation (planes, trucks, automobiles, trains and ships), structural collapse (buildings, dams, bridges, mines, and other structures), explosions, fires, chemical (toxic waste and pollution), and biological (sanitation).

In addition to HVA, it is critical that the program should provide for staff training, locating a designated command center, listing of personal hygiene supplies for working staff, and completion of systems/devices vulnerability assessment. The assessment should include identification of possible dependencies between systems, that is, impacts from the loss of electricity on the operation of the dialysis water system, or from the loss of routers and wireless access points are important types of information that is critical during a disaster. Evaluating the dependency of one system on another and the probability and severity that such events can occur are important assessment tools. Other vulnerabilities include bedside monitoring systems which are dependent on switches, servers, and the like that are powered by commercial power without emergency system backup or hand-held radios that are dependent on active antennas for communication, paging systems, or wireless phones of which the antenna is prone to wind damage or not supported by emergency power. The central vacuum or pressured gas systems should be identified and evaluated regarding the anticipated impact on patient care if, for example, commercial electrical or emergency power is lost. The assessment also should provide information on the age and condition of the batteries in critical equipment such as portable monitors and infusion pumps. If the batteries are close to their end of life, establish a monitoring and maintenance program to keep them reliably refreshed before disaster strikes. Always have backup plans and supplies available. Designation of critical medical equipment and systems should be completed prior to disaster events. Critical systems and patient critical equipment should be listed on the hazard vulnerability table, along with assessment and prioritization and triaging plans, making sure that these assessments are all communicated to the clinical engineering program. The clinical engineering program, using such information, will staff personnel who are skilled in these systems and equipment as part of their disaster preparedness planning.

Before disaster strikes, consideration is needed for assessing how to safely shut down systems requiring stoppage. With pre-scripted rapid shutdown protocols as time permits, bringing systems down in a scheduled manner will prevent damage, and prevent impact on other systems and loss of data, for example a sophisticated analyzer in the laboratory or a water purification system in the dialysis area. Systems are sensitive to variation in water quality or temperatures and you are better off shutting them down. An inventory list with the location of vulnerable systems, disaster protocol, and a time estimate for shutting them down are necessary to schedule the successful initiation of this activity. Before disaster strikes, laboratory, radiology, research, and dialysis types of equipment should have a list of how/when shut down is needed.

Being prepared for a disaster makes the difference. Clinical engineering managers should designate personnel to staff the three stages of disaster event. Managers should assign their staff into teams A, B, or C. Team A, the pre-disaster or preparation team. Team B rides out the event; has to be prepared to ride it out for a period of up to several days of the emergency situation in shifts. Staff should have personal items including: sleeping bag, pillow, blankets, several changes of clothes, particularly undergarments, toiletries (toothpaste, shampoo, deodorant, and contact lens), extra flashlights, bottles of water, medication supplies, books, playing cards and games, cellular telephones, and charging units. It's a tense situation and staff will need breaks to avoid exhaustion.

Some radiological equipment at hospitals emit radiation, such as x-ray units, CT scanners, and linear accelerators. Others contain sealed radionuclides used in teletherapy and brachytherapy devices, and others work with unsealed radiopharmaceuticals such as gamma cameras, Single-Photon Emission Computed Tomography (SPECT), and Positron Emission Tomography (PET) units. All these devices present special radiation safety issues during a disaster. Yet, their availability may be critical for assessing victims' prognoses and, thus, performing adequate triage. Medical and health physicists in a hospital or clinic must be prepared to cope with contingencies such as the source of a Cobalt-60 unit becoming stuck in the unshielded position or the rupture of a vial containing radioactivity in a nuclear medicine department. To be ready, Medical Physicists/Radiation Protection Officers need to have contingency plans and do periodic drills to test the appropriateness of the responses. From the design point of view, medical devices containing radioactive sources must have a manual retract assembly and/or a UPS to prevent patient/staff irradiation in case of a power failure (i.e., source would not retract).

Hospital physicists must also be prepared in case of a nuclear or radiological disaster (e.g., Chernobyl) to liaise with the authorities dealing with response and recovery operations. The effects of disasters may be severe. Medical services may be inoperable if radiation contamination is serious and both patients and staff may have to be relocated. In such cases, hospital managers should request help from the national agency in charge of disaster response, and the country may need international assistance (IAEA, 2013). If the hospital is part of the national/regional network of hospitals providing medical care to irradiated or contaminated patients in a nuclear/radiological emergency, and if regulatory authorities have estimated that its own level of contamination is acceptable, the hospital should immediately take the following actions: (i) Activate the Emergency Plan, that should have been tested in practice drills periodically, including coordination with the National Disaster Response Agency, (ii) Assemble medical/technical/radiation expert teams, and (iii) Prepare the hospital to provide rooms or areas for: (a) irradiated patients in need of sterile conditions, (b) radioactivity detection in incoming patients and staff, and (c) decontamination.

The role of medical physicists in natural or man-made disasters is quite different. They should ensure that devices and buildings are built to withstand major potential disasters in the area. If the event occurs when patients are undergoing radiological procedures, the hospital staff should stop exams, interventions or treatments, move the patients to a safe location and in the case of radiotherapy treatments, record the given doses (in monitor units or time). After the event, medical physicists need to check and correct—if possible—the mechanical and electrical integrity of the medical device's components, their accessories, including patients' masks, immobilizers, etc., and of dosimetry systems and quality assurance phantoms. They should ensure that the device's electrical and water supplies are working or have them repaired, if needed. They should also verify the operability of software and networks.

Before returning to clinical use, all radiological equipment must be fully evaluated and recalibrated by a medical physicist. If the device contains a radioactive source, the radiation protection officer has to verify that the source is still in the device or container, that the sealed source encapsulation is intact and that either no contamination has occurred or the contaminated areas have been decontaminated. In the case of earthquakes, there are additional considerations, since, depending on their magnitude, earthquakes may affect the alignment of the radiation beam. After the event, for medical imaging devices, the medical physicist has to check the congruence of the radiation and light fields and the alignment of the whole imaging chain, including displays and networks (RIS, PACS). For external beam radiotherapy devices, especially linear accelerators, he/she has to check: (i) The position of collimator, gantry and table isocenters, (ii) Field flatness and symmetry, (iii) All dosimetry and treatment planning systems, (iv) In-room imaging devices (IGRT), (v) The record and verify network, (vi) All patient accessories.

In conclusion, devices used for medical imaging and radiation therapy are very vulnerable to disasters due to their design complexity. Prevention and response measures should take into account the medical device itself and its role in the whole radiological process.

Levels of Responsibility

There is a robust network of organizational capabilities that have been developed over decades to provide preparation and support for disaster management. These programs are organized at different levels to coordinate efforts and resources, from international to local hospital levels.

INTERNATIONAL LEVEL: The United Nations provides an integrated organizational framework for nation-states to identify resources and plan for disaster preparedness and response. The multiple agencies involved are shown in Figure 11.4. In case of nuclear/radiological disasters, other agencies are also involved. It may be challenging for clinical engineers to identify the specific technology-related plan. Some agencies coordinate information with the WHO, but they also may provide systems for communication and collaboration with other external organizations operating in the disaster preparedness domain. Other international organizations, such as The International Federation of Red Cross and Red Crescent Societies, the International Committee of the Red Cross, CARE, Medecins sans Frontières, and the International Rescue Committee provide valuable leadership, training, and rapid-response capability for disaster management.

FIGURE 11.4 United Nations disaster agencies. *Source: Unocha.org*

NATIONAL LEVEL: Each nation will have different degrees of organization and resources dedicated to disaster preparedness and management. In the United States, the Federal Emergency Management Agency (FEMA) is designated as the primary emergency preparedness and response institution for the nation. It operates an Emergency Management Institute to train officials at all levels of government http://training.fema.gov/emi.aspx) and to run disaster exercises. The FEMA Incident Command organization is structured as follows (Figure 11.5).

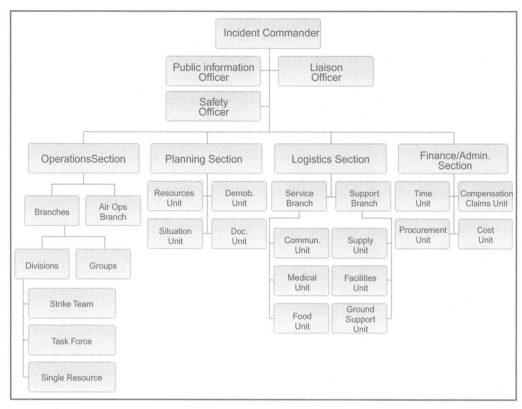

FIGURE 11.5 FEMA incident command organization template. *Source: http://emilms.fema.gov/IS200b/assets/ ics200hc_l02_s30.gif*

STATE LEVEL: A State-level Emergency Operations Center might be organized as shown in Figure 11.6. Note how the State structure duplicates many of the functions of the national model. This enables everyone to use common vocabularies and procedures to manage disaster situations.

CITY LEVEL: A local fire department can also serve as an Emergency Services Department whose duties include:

- **Preparing** the city by maintaining the city's Emergency Plan and Emergency Operations Center (EOC), and designing and conducting drills and exercises.
- **Training** all city staff on the Standardized Emergency Management System (SEMS) and personal preparedness, as well as recruiting and training members of the city Emergency Response Team (ERT).
- **Planning and coordinating** response for emergencies with other local jurisdictions and Regional, State, and Federal Agencies is also facilitated by OES.

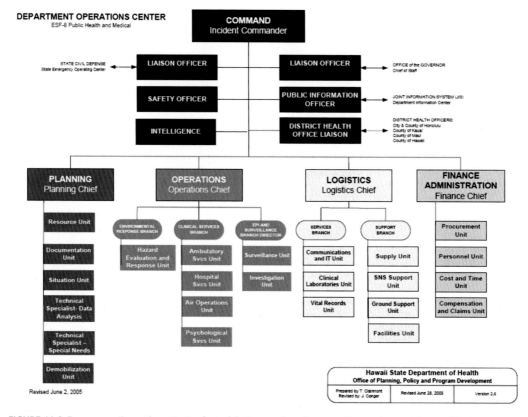

FIGURE 11.6 Emergency Operations Center (sample). *Source: http://www.ualbanycphp.org/pinata/phics/images/fig03.gif*

- **Serving as a resource** for schools, businesses, community groups, service organizations, and neighborhood associations, providing information, training, assisting with exercises, and participating in community events.
- Conducting Community Emergency Response Team (CERT) **training** to prepare individuals to assist in disaster recovery on teams and with neighborhood groups. Training and certification is offered to individuals, organizations, school districts, companies, and neighborhood groups.
- Assisting in the **formation of Neighborhood CERT Groups** which support the CERT program with information to residents and periodic exercises to practice and maintain CERT skills. When 911 services are not available in a disaster, neighborhood groups can be critical in collecting and relaying neighborhood needs to the city's Emergency Operating Center.

    Source: http://www.ci.mtnview.ca.us/city_hall/fire/programs_n_services/disaster_preparedness.asp

United Nations Office for Disaster Risk Reduction (UNISDR) has developed a "Ten-point Checklist: Essentials for Making Cities Resilient" that provides a useful reference for understanding where your organization and others may fit into the city's disaster management picture.

Source: http://www.unisdr.org/campaign/resilientcities/toolkit/essentials

HOSPITAL LEVEL: A detailed California program for hospital incident command center can be found at http://www.emsa.ca.gov/disaster_medical_services_division_hospital_incident_command_system

In sum, disaster management offers biomedical and clinical engineering professions a valuable chance to update and expand their scopes of practice, their skill sets, their training, their service models, and their leadership roles in the DM systems available to them.

Expanding the Career Path

Clinical Engineers (CEs) already face professional challenges on many fronts: organizational capacity, public expectations, 24×7 service models, constant technological change, and economic constraints, among others. Regional and local infrastructures vary significantly across the globe, and disasters often have impacts that cross official jurisdictional boundaries in ways that create immediate uncertainty regarding who is responsible for what, and what the resources are that can be deployed for rescue and recovery. We provide an overview here of some resources that can help CEs expand and clarify their roles for disaster preparedness and response. The goal is to define clinical engineers as knowledgeable, visible, and capable professional partners in disaster management policy, plans, budgets, exercises, and specialized coordination of response and recovery efforts.

You Don't Have to Re-invent the Wheel

Many of the existing DM roles are already highly professionalized and staffed by people with decades of experience, so it is important to understand what skills and tools have already been formally tested and implemented, to avoid duplication of effort or conflicting views of responsibility. Some formal investigation will be necessary to track down existing plans and Mutual Aid agreements (or to verify their absence) and to identify gaps and overlaps, so that CEs can customize their roles according to local circumstances.

Numerous Causes of Disaster Require Clinical Awareness of Community Systems

Clinical engineers can play a useful role in helping their own organizations and community members understand how a disaster can have multiple causes, contributing factors, and outcomes, many of which are directly under human control and which therefore permit plans to mitigate the damages. Some of these causes can be managed in advance (such as through the adoption and enforcement of building construction standards), to reduce/mitigate the impact of a disaster. Moreover, some disasters may themselves cause secondary disasters that need to be anticipated and mitigated in advance wherever possible. The more CEs understand the complex network of causal factors and constraints, the more they can have a positive influence on those factors by working through community and government agencies, local hospitals regional stakeholder groups to be prepared on multiple fronts. Increasing emphasis on mitigation and resilience is an important approach to help communities.

Clinical engineers can bring valuable planning and operational expertise to the discussions and can offer new technical resources to strengthen disaster management capacities. Causal factors to discuss include:

• Nature, weather, geology	• Public ignorance, apathy, lack of information
• False expertise	• Technology failure
• Optimistic/faulty assumptions	• Hostility (ethnic, political, economic, religious, national)
• Deception; denial of information	• Failure to prepare
• Failure to respond	• Faulty maintenance procedures
• Human error	• Inadequate physical infrastructure and human resources
• Economic fraud, criminality	• Terrorist acts
	• Political motivations and corruption

Training Resources

It is important that CEs have basic training in emergency/disaster response, and there are many organizations that can provide the foundations. In the United States, the Community Emergency Response Team (CERT) Program educates people about disaster preparedness for hazards that may impact their area and trains them in basic disaster response skills, such as fire safety, light search and rescue, team organization, and disaster medical operations. Using the training learned in the classroom and during exercises, CERT members can assist others in their neighborhood or workplace following an event when professional responders are not immediately available to help. CERT members also are encouraged to support emergency response agencies by taking a more active role in emergency preparedness projects in their community. CEs might initiate CERT training programs in their areas, to create a network of future collaborators during disasters, and to establish a minimum framework of understanding for procedures, roles, and expectations. CEs can also play a role to promote remediation activities and build resilience into their communities.

Source: http://www.citizencorps.gov/cert/index.shtm

Liaison to Stakeholders

CEs can play a valuable role in professional networking, education, and developing special response procedures for the different stakeholders that may exist in their community. There may be many stakeholder groups whose disaster needs may differ, depending on timing and location of disaster, and their possible roles in mitigation and response. These stakeholder groups are explained in Table 11.1.

Table 11.1   Stakeholder Groups to Consult for Disaster Preparedness and Response

Building Science Professionals	Language translators
Contractors and Vendors	Livestock Owners
Children's Working Group	Parents and Teachers
Disaster Survivors	Pet Owners
Emergency Managers	Individuals with Special Needs (mental or
and Personnel	physical disability, dialysis, different languages)
Fire Service	Press Resources
Government (Federal, local, and state)	Private Sector
HAZUS User Groups	Tribal Representatives
Home (Property) Owners	Universities
Individuals	Volunteers
Institutions	Maternity and neonatal patients
Prison populations	
People without vehicles	
Children	

Readiness Assessment

Clinical engineers should work closely with their facility and community disaster preparedness leaders to identify existing resources, procedures, and Mutual Assistance Agreements, and evaluate how new technologies might be adopted to improve preparedness and responsiveness. Depending on the location and type of disaster, the existing resources and jurisdictions may be very different. For example, a bioterror disaster event in a major city in the United States will quickly involve State and Federal agencies and military organizations that have complex, predefined roles, responsibilities, and resources to use, whereas a tsunami in Southeast Asia may affect many nations who have little disaster-specific capabilities other than using military helicopters to ferry donated supplies to affected areas. The role of CEs therefore will vary depending on the institutional and financial infrastructures of their locations. It is particularly important to understand how readiness and response capabilities and roles will differ depending on the scale of the disaster. An emergency may be successfully managed by a local facility or city agency, whereas a disaster, or mass-casualty disaster, will require more sophisticated planning and pre-defined agreements on resource controls and decision-making. Take advantage of methods developed by WHO to help hospitals review their readiness for emergencies and disasters, such as: http://www.unisdr.org/.

Source: http://www.euro.who.int/__data/assets/pdf_file/0020/148214/e95978.pdf?ua=1

New Tools, New Opportunities

Recent advances in software, hardware, and telecommunications have made it possible for communities to be much better prepared and informed during disasters, to help direct scarce resources where they are most needed. The following programs offer significant advances over earlier capabilities for enabling highly informed, real-time responsiveness to

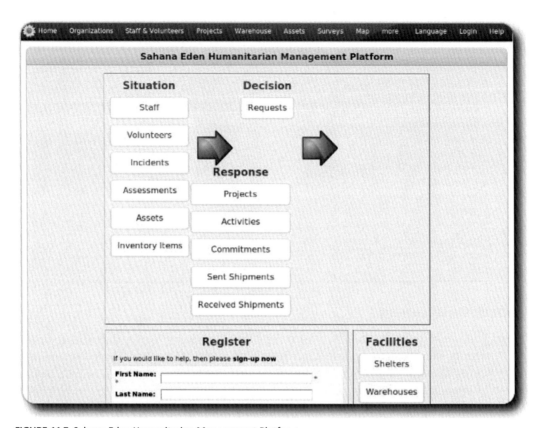

FIGURE 11.7 Sahana Eden Humanitarian Management Platform.

disaster conditions. Many of these capabilities focus on "early situational awareness" so that first responders and institutions can have timely, accurate understanding of the damage conditions in the region.

- **Sahana Foundation** software (Figure 11.7) provides two rich functionality platforms (Eden and Vesuvius) for humanitarian relief and medical response activities, enabling users to track Situations, Decisions, and Responses for different program activity levels, including Incidents, Assessments, Person Search, Assets, Inventory, Requests, Projects, Commitments, Sent Shipments, and Received Shipments. Users can keep track of Organizations, Staff and Volunteers, Surveys, and Maps that pinpoint local conditions.

  http://sahanafoundation.org/products/

- **Ushahidi** is a free, open source platform that enables users to visualize various kinds of disaster information on maps, to track reports, filter data by time, and collect information via text messages, email, twitter, and web-forms.
  Source: (http://www.ushahidi.com/product/ushahidi/)

- Common Operating Picture (COP) is a generic platform concept that takes different forms depending on the customized requirements of the end-user organization, who is typically a first responder entity such as a fire department, ambulance service, police department, or public health agency. COP supports rich pre-population of data from organizations that can be displayed in clickable layers on Google Maps, providing rapid access to people, places, things, and events that are relevant to disaster preparedness and response. Schools, hospitals, fire hydrants, gas lines, and other "fixed assets" have their own distinctive icons, and mobile assets such as fire trucks, ambulances, rescue units, and police cars can be tracked in real time. Smart phones can download specialized Disaster Response applications that enable citizens to phone in site-specific information about disaster conditions, including text, photos, video, voice messages, and brief forms that classify the conditions (fire, damaged freeway, flood, explosion, etc.). Advanced models can include detailed floor plans for critical infrastructure buildings, to enable detailed damage assessments and other site information that can be instantly communicated to other local and regional COP users via clickable map icons that open to display predefined information categories and links to related resources.

  Recent refinements of the COP have been developed by David Coggeshall, of California Communications Inc. (http://comopview.org/sfc/) for the Golden Gate Safety Network of the San Francisco, California Bay Area, building on state-of-the-art work done in the multi-year Carnegie Mellon Silicon Valley Disaster Management Initiative (http://www.cmu.edu/silicon-valley/dmi/). This software enables the pre-loading of zoomable floor- and room-specific architectural drawings for critical infrastructure, such as schools, hospitals, large office buildings, and public utilities. This enables invaluable capability for real-time damage assessment and rescue prioritization that combines precisely localized photos, audio commentary, and use of standardized icons to represent essential infrastructure elements (Figure 11.8).

- **Hastily Formed Network** in a Suitcase—HFN is a portable, suitcase-size ensemble of technology that contains alternate power, completely self-contained communications, and a back office in a box with virtual servers, applications and desktops using VMWare. It allows users to replicate office servers, Microsoft Office applications, Outlook e-mail, calendars, your 911 call system, and other back office applications. This enables emergency/disaster responders and managers to have a rapid, mobile capability for maintaining critical communications, if local conditions require rapid re-location.

  Source: http://nps.edu/Academics/Institutes/Cebrowski/HFN.html

- **Google Crisis Response**—Google offers several online tools that can help communities manage alerts, find mission persons, localize disaster conditions on maps, This platform enables users to:
  - ISSUE ALERTS: "During a crisis, individuals go online to search for the latest emergency information. Google has created a platform to disseminate relevant emergency alerts to users when and where they're searching for them. Public Alerts

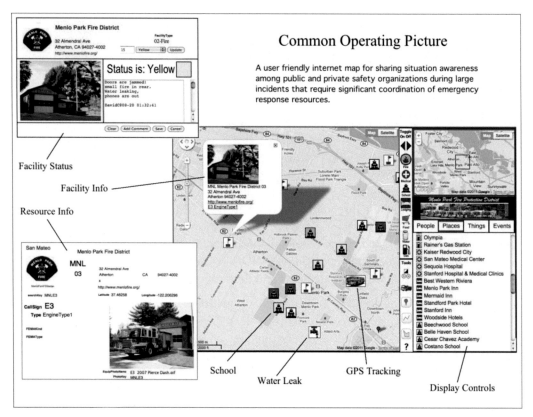

FIGURE 11.8 Common operating picture. *Source: http://comopview.org/sfc/*

are available on Search, Maps and Google Now You can use Public Alerts to get your information to the public."

- LOCATE PEOPLE: "Following a crisis, people often get separated, and responders play a role in helping people locate one another. Google Person Finder, launched by the Google Crisis Response team, helps with this process by providing an open platform for individuals and organizations to let people know who they're looking for and to enter updates about missing persons. As an organization you can: Embed Google Person Finder in your website to allow people to directly access and use the tool; download data from Google Person Finder to match with your information or take to the field; upload data you've collected into Google Person Finder."

- DO CRISIS MAPPING: "In a crisis, users have diverse information needs and limited time to learn new tools. Crisis Map, a mashup tool built on the familiar Google Maps API, aims to put critical disaster-related geographic data in context, and in a

map-based viewing frame optimized for usability across a range of browsers and mobile devices. With Crisis Map you can do the following:
- Explore disaster-related geographic data without any special software or GIS expertise required
- Embed a Google Crisis Map on your website
- Share a crisis map with co-workers, media outlets, and partners
- Interact and download data from a crisis map
- Contribute data to be included on a crisis map."
• CREATE FUSION TABLES: Users can "gather, visualize and share data online with your staff and other response organizations and constituents to:
  - Visualize your data from shelter lists to power outages instantly as a map or a chart.
  - Identify data patterns to aid in crisis decision making
  - Show the world your work in real time by embedding your map or chart in a web page
  - Collaborate with other responders by merging your data, allowing you to see all important related information in one place."

Include Low-Tech Options for Communication

Clinical engineers can be trained and licensed to use Ham radios, to provide emergency capability for their facilities, if cellular and land lines are damaged.

Alternate Sites of Care

Clinical engineers may be involved in planning, implementing, and operating medical services in locations other than the primary hospital or clinic where they work during normal circumstances. Disasters and refugee populations may generate medical needs that exceed normal hospital capacities, and it may become necessary to transfer or re-route patients who are injured to other treatment locations, such as schools, stadiums, or rapidly deployable facilities. Clinical engineers will be needed to plan these capabilities and logistics in advance of disasters, and ensure compatibility of consumables (such as infusion pump tubing), adequacy of power supplies, and appropriate integration of medical devices with provisional informatics systems (Figures 11.9 and 11.10).

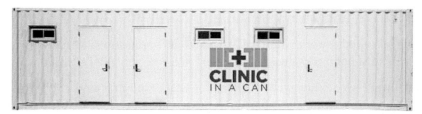

FIGURE 11.9 Industrial shipping containers can be converted into clinics. *Source: http://www.clinicinacan.org/*

FIGURE 11.10 ICRC Rapid Deployment Emergency Hospital. *Source: http://procurement.ifrc.org/catalogue/*

Emergency Electronic Medical Records

For hospitals that do not have electronic medical records, various software programs exist that maintain care-critical information about a patient's medical history in a database that can be accessed by Emergency Medical Technicians and hospital clinicians to avoid medical errors, notify loved ones, and inform responders about any allergies, medications, medical conditions, and advanced directive information that may affect treatment. One example is http://www.mye-mergency.com/.

Emergency Identity Management for Staff, Patients, Volunteers, and Movable Equipment

Specialized hand-held devices have been developed to help disaster managers keep track of people and resources under very dynamic disaster conditions. Since many normal hospital operations (such as maintenance services) may be temporarily suspended during disasters, clinical and biomedical engineers may have very different roles to play. Since CEs may already have experience in clinical procedures involving patient and clinician identification, change management, and asset tracking, these skills can be quickly put to work in disaster events where security needs are heightened, but normal safeguards and controls may be damaged or lacking. CEs may help fill the gap by with special technology for ID tagging of patients and staff and medical devices that are being re-located to alternate sites of care. An example is at http://vanirtech.com/solutions.

Other Skill Areas and Responsibilities

Clinical engineers should work with local, regional and professional leadership to upgrade job descriptions to support the special operational needs of a disaster situation, including:

- Managing local/regional Emergency Operations Center information as Situation Specialist
  - Filter and communicate incoming info for Common Operating Picture
- Diagnosis and remediation of wireless infrastructure in local and alternate sites of care

- Conduct damage assessment rounds to validate operational state of medical device infrastructure
- Workflow engineering for alternate sites of care
  - Remote patient monitoring
  - Mobility workflows for disaster situations
- Ham radio services; supplementary communications channels
- Alternate supply-chain management
  - Credentialing and access control to sensitive areas
  - Patient and staff identification, tracking technologies linked to vital signs monitoring
- Extension of emergency infrastructure: power, device transfer, connectivity to remote sites
- Register and manage volunteer teams for disaster response
- Manage surge-site setup.

Cooperation with Other Professionals

It is important for the success of emergency plans that good cooperative relationships be established between CEs and other professionals such as medical physicists and radiation safety officers, especially when dealing with radiological or radioactive materials in hospitals. To ensure coordination at the response stage, the cooperation should start during the preparation of the emergency plan, where roles and responsibilities are clearly established, and tried out in the consequent drills and exercises. Regularly scheduled drills are an important part of skill development and establishing a positive sense of teamwork for difficult circumstances.

Hospital staff should be prepared to cope with disasters through frequent drills of a well-developed emergency plan that encompasses the phases before, during, and after the disaster, and that includes radiation protection considerations. Remember, there may be scenarios that you did not think of or plan for, however, these tools will assist you to manage such situations that may seem to be too devastating or disastrous to overcome. With planning and training, you will overcome them. The rewards in lives saved will be immense.

Concluding Advice

The faster clinical engineers embrace the expanded program, tools, and skills detailed here, the sooner they will be able to move into positions of partnership and leadership, to develop new organizational capabilities that will save lives and reduce suffering. There may never be a perfect preparation for disasters, but we are fortunate to live in a time when technology and clinical engineering professionalism are at a high point that can transform weak healthcare systems into robust ones. This will require clinical engineers to step up enthusiastically into these new roles, to help create "horizontal intelligence" in their healthcare systems that will enable the "vertical intelligences" of their organizations to cohere under challenging circumstances.

Clinical engineers and their clinical and facility peers should conduct formal drills regularly, and incorporate disaster management tools and processes into their normal daily activities to increase their skill and readiness. Although these exercises are important, it is important not to just plan for small, local events that you can manage successfully. You should "Test to Fail" with truly stressful, regional scenarios so you can discover weaknesses in your program and in your wider system relations.

- Make contingency plans, in case that designated critical equipment or system fails, or staff members are absent (could not commute to their job), or communication systems are down, and the disaster occurs at inconvenient times and places.
- Actively investigate the potential role of military personnel in large-scale disasters and establish liaison relationships in advance of actual disasters. Explore issues of supply-chain coordination, compatibility of infusion and pharmaceutical procedures and supplies, and compatibility of emergency electronic records and other vital sign monitoring technologies.
- Have a communication plan ready for guiding your program during disaster that includes the relocation of your laboratory or other essential functions.
- Do not ignore the need of rest space, food, medications your staff may need, and personal hygiene.
- Provide positive role modeling to others under stressful circumstances and help build a collaborative, mutually supportive professional culture for disaster management.

With a good plan and practice, your clinical engineering program stands to make a positive difference at the most demanding and challenged time.

References

Businessdictionary. 2014. Adopted from: http://www.businessdictionary.com/definition/disaster-preparedness.html (accessed 29.04.14).

CRED, 2012. *Annual Disaster Statistical Review 2012. The Numbers and Trends*, WHO Collaborating Centre for Research on Epidemiology of Disasters — CRED. Available at: http://www.cred.be/sites/default/files/ADSR_2012.pdf (accessed 26.04.14).

David, Y., Borrás, C., Hosea, F. 2013. Disaster Preparedness Program for Health Facility's Technology Managers, WHO 2nd Global Forum on Medical Devices, Geneva, Switzerland.

David, Y., Dreps, D. 2006. Clinical engineering role in the emergency preparedness plan. ACCE education series telecast, December 2006, www.accenet.org.

Disaster Resilience, 2012. A National Imperative, Committee on Increasing National Resilience to Hazards and Disasters Committee on Science, Engineering, and Public Policy. The National Academies, The National Academies Press, Washington, D.C.

http://eden.sahanafoundation.org/.

International Atomic Energy Agency. 2013. Radiation Emergency Management Plan of the International Organizations EPR-JPLAN IAEA, Vienna.

Joint Commission. 2006. Standing Together — An emergency planning guide for America's communities, Joint Commission. http://www.jointcommission.org/assets/1/18/planning_guide.pdf (accessed 18.05.15).

Joint Commission. 2008. Emergency Management, The Joint Commission Standards Information. http://www.jointcommission.org/standards_information/jcfaqdetails.aspx?StandardsFaqId=521& ProgramId=47 (accessed 18.05.15).

Joint Commission. 2013. New and Revised Requirements Address Emergency Management Oversight, Hospital manual. Joint Commission Perspective, 33(7). http://www.jointcommission.org/assets/1/18/JCP0713_Emergency_Mgmt_Oversight.pdf (accessed 18.05.15).

Leaning, J., Guha-Sapir, D., 2013. Natural disasters, armed conflict, and public health. N. Engl. J. Med.

Other Resources

United Nations Office For Disaster Risk Reduction - http://www.unisdr.org/

"The mandate of UNISDR expanded in 2001 to serve as the focal point in the United Nations system for the coordination of disaster reduction and to ensure synergies among the disaster reduction activities of the United Nations system and regional organizations and activities in socio-economic and humanitarian fields."

Federal Emergency Management Agency.

In the United States, FEMA has the lead role in managing disaster preparedness and response. http://www.fema.gov/emergency/nims/

International Association of Emergency Managers - http://www.iaem.com/

IAEM promotes the "Principles of Emergency Management" and represents professionals whose goals are saving lives and protecting property and the environment during emergencies and disasters. Its mission is to provide information, networking and professional opportunities, and to advance the emergency management profession.

Response Systems – Disaster Preparedness for Healthcare and Communities, JCAHO, http://www.disaster-preparation.net/resources.html (last accessed May 18, 2015).

Ambulance Sciences (https://www.jsomonline.org/CareerCenter.html) can provide a foundation of academic and certificate topics centered on the role of ambulance technicians. This can be a starting point to model some of the skills, resources, and roles in your area that could be organized for disaster response. The equipping of ambulances with telecommunication capabilities can be a valuable telemedicine resource for both normal and disaster conditions, so that vital signs and other clinical information (such as the patient's electronic medical record) can be mutually accessible to hospital and ambulance staff, enabling earlier and more accurate treatment interventions.

University of Delaware, Disaster Research Center - http://www.udel.edu/DRC/

University of California, Davis - Point of Care Technologies Center - http://www.ucdmc.ucdavis.edu/pathology/poctcenter/

12

Human Factors Engineering in Healthcare

Torsten Gruchmann
USE-LAB GMBH, STEINFURT, GERMANY

Human Factor Engineering for Medical Devices

Introduction

As medical technologies become more and more advanced, corresponding *patient safety* concerns are also growing in number and intensity. The media has also become interested in the topic: Something goes wrong at the hospital, a patient is injured or killed; the public wants answers. These answers are often associated with discussions of physician errors, how could the doctor make such a grievous mistake? But often this assumption of *user-error*, a clinician did something wrong, is wrong itself; the problem actually lies in a *use error*. While a user-error is mono causal, the error is based on a single cause, a use error is poly causal triggered by the user or the device.

Use errors are defined as actions or omissions of actions that lead to (medical) device responses other than those intended by the manufacturer or expected by the user.

When accidents happen, whether by user-error or use error, the patient is usually the one to suffer.

Imagine you've gotten really sick. For the sake of your health, and in hopes of getting better soon, you seek out a hospital and leave your fate in the hands of the experts.

What you aren't expecting—and what you certainly aren't hoping for—is a drug overdose or a hospital infection because instruments weren't kept properly sterile. You aren't thinking about electrodes that will deliver a too strong current or about how maybe no one will notice the alarm of the life-supporting device you're hooked up to going off. You probably haven't wondered if the hospital bed you're lying in might become a death trap.

Unfortunately, these many things you don't want, expect or even think about happen all too often. Researchers at the Institute of Medicine in the United Sates published an article in 1999 called "To Err is Human" (Institute of Medicine, 2000); in it they report that every year somewhere from 44,000 to 98,000 people die in the United States following medical errors. More recently, a study in Germany (AOK) cited 18,000 avoidable deaths in 2011 following *medical errors*. Putting this number into perspective, patients were treated 18.8 million times that year; that means every 1000th patient died as a direct result of their treatment. Compare this with 4000 automobile-related deaths in Germany or 500 yearly plane-crash mortalities worldwide on average. And yet, while the media immediately and constantly cover plane crashes there is scant reporting about the 18,800 people who died following medical errors in hospitals—the equivalent of about 35 Airbus A380 crashes!

This is not to say that the matter is receiving no attention from lawmakers. US Senator Bernie Sanders called attention to the 440,000 US deaths because of preventable medical errors in hospitals a year in 2014 (Sanders). Quoting the *Journal of Patient Safety* (James, 2013), he pointed out that this made such errors the third leading cause of deaths in the United States.

The fact that any one of us could, at any time, suddenly become one of the unfortunate patient statistics makes this topic one of enormous importance and general interest. Throughout this discussion, we must not forget that millions of seriously injured and ill people are treated successfully every year and that life expectancies are increasing from decade to decade due to better medical care.

Stress as a Cause of Use Error

Looking at the question of why—sometimes life-threatening—mistakes are made, we find a primary culprit in the ever-increasing psychological and physiological pressures upon nurses and physicians. This pressure is a necessary consequence of the simultaneous increase in cases and decrease in budget alongside a shrinking number of qualified personnel. Not to be forgotten at this point is the growing competition that is affecting even hospitals now. Unfortunately, these, too, are matters that as yet receive little attention from the public.

Turning again to one of the most common sources of medical mistakes, we must first define *stress*. Organisms respond to so-called *stressors*—external or internal challenges, like the constant requirement see a large number of patients in a short time—with a combination of physiological and psychological responses. While these responses on the one hand allow individuals to meet challenges, they bring with them, on the other hand, physiological (e.g., tiredness) and psychological (e.g., irritability) consequences. *Stress*, then, encompasses both the enabling and the disabling aspects of the reaction.

Much of the stress experienced by professionals in the healthcare industry can be attributed to an ever-growing workload that takes a bodily toll and also leads to internal stressors like a sense of decreased control. Additional stressors come from without, pressure placed on the individual by colleagues and patients.

This all leads to the unhappy state of affairs in which medical professionals can no longer truly care for patients, but rather simply function themselves in a machine-like manner. And it is in this state of affairs that nurses and physicians must also operate numerous medical devices, more than 100 different ones for those working in intensive care. Given the number, diversity, and sometimes highly complex nature of the technology that must be used by individuals under great stress, it should surprise no one that mistakes are made.

Design Deficiencies in Medical Devices as a Cause of Incidents

Above, the diversity of the medical devices that must be operated in clinical facilities was brought up; not only are the devices themselves diverse in aim, functionality and build, the user interfaces also vary considerably. At this point, it is important to note that *user interface* always refers to all parts of a device or system that a person interacts with, including hardware as well as software interfaces and not just, for example, the display commonly called *graphical user interface*. Additionally, all accompanying documents are per definition part of the user interface.

Particularly in stressful situations, inadequate designs can lead to incorrect use. Design-based sources of error are numerous and can lead to lessened safety of use, but also to decreased user-friendliness; a sample of *possible sources of error* is given here:

- Illogical, awkward action-sequences
- Missing visual, tactile or audible feedback
- Insufficient user-guidance and status-information
- Unclear setbacks/requirements
- Hidden or confusing functions
- Default parameters that do not match user expectations
- Unclear symbols, captions, or codes
- Unfamiliar terminology
- Inadequate grouping
- Inadequate safety measures
- Inconsequent formats
- Inadequate user manuals
- Unclear labeling of products and packaging
- Inadequate consideration of intercultural differences.

For an illustration of the user-unfriendliness of many medical device designs, visit a clinic and count how many times you see tape or bandages used in unconventional ways. You're likely to find it wrapped around gauze that's placed over sharp corners, fastening one device to another, and provided any number of labels that the manufacturer did not (Figure 12.1).

Normative Rules as a Method of Risk Reduction

Today there are strong demands on medical device's user interfaces: They need to be easy, intuitive, and safe to use—even in stressful situations. These demands, intended to prevent accidents caused by miscommunication between man and machine, are legally binding for medical device technology. The *European Medical Device Directive* (MDD 93/42/EWC) with its amendment from 2007 (2007/47/EC) counts ergonomic considerations for patient safety among its essential requirements. To fulfill these requirements published by the European Union, guidelines for the application of usability principles are provided in the international standard *IEC* 62366-1:2015 "Medical devices—Part 1: Application of usability engineering to

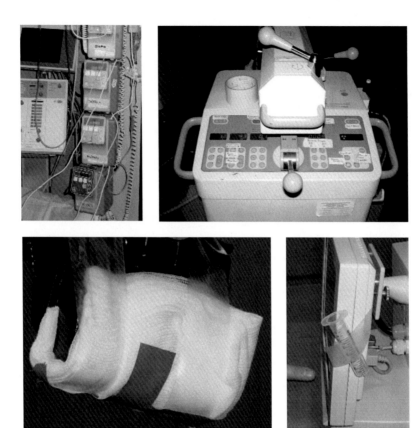

FIGURE 12.1 Tape used to improve safety.

medical devices." This standard is applicable to all medical devices disregarding if it is an active or non-active medical device.

The IEC 62366-1:2015 as an international recognized consensus standard in combination with the FDA human factors guidance papers is also applicable to fulfill the *usability engineering* requirements of the Code of Federal regulations, monitored for compliance by the FDA.

To reduce the risk for a medical error and improve patient safety, there is a very simple recipe with two ingredients. Manufacturers can deliver the first ingredient by implementing and faithfully following the usability oriented development process defined in IEC 62366-1:2015 and the demands it makes regarding the analysis, specification, development, and evaluation of this process. The second ingredient depends on law- and healthcare decision-makers working in concert with medical professionals to create a culture of openness regarding mistakes, for example, within the framework of a *Critical Incident Reporting System* (CIRS).

The following section is discussing in detail what the usability engineering process, claimed by the IEC 62366-1:2015, is asking for.

A Usability Engineering Process following IEC 62366-1:2015

The previous section elucidated the regulatory background of the IEC 62366-1:2015; the following chapter is about the process the standard defines and what has to be documented throughout the medical device development.

The IEC 62366-1:2015 is an international standard that describes a usability engineering process and provides guidance on how to implement and execute the process so that only acceptable risks related to the usability of a medical device remains. The usability engineering process is of iterative nature and can be easily combined with the design and development process leading to the *user-centered design* or *usable design process*. Figure 12.2 shows the links between the design and the usability engineering process portrayed on a simple illustration of a product development process.

The usability engineering process is intended to identify and minimize use errors and thereby reduce use-associated risks already in an early development phase. As described earlier, use errors can lead to hazardous situations in which humans get hurt.

In this context, a very frequent question to be answered is about the difference *between usability engineering* and *human factors engineering.* Some experts use the terms "human factors engineering" (HFE) and "usability engineering" synonymously, while others differentiate between them. The latter perceive HFE to be, above all, the exploration of the user and the

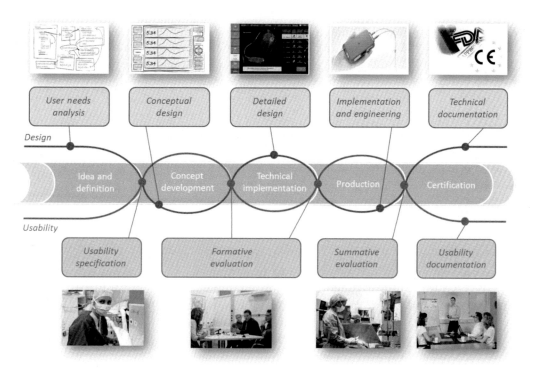

FIGURE 12.2 The user-centered design or usable design process.

actual user interface design. Usability engineering is then considered to mean the evaluation of a user interface via usability objectives and usability tests. Regardless of the terminology, both methods—and many similar methods—lead to the enhanced usability of a product. The central idea behind HFE is that common sense is usually not enough to create a user-friendly product. User-friendliness is rather the result of applying HFE in the product development process.

Definition of the Use Specification

One of the most important principles is to start Usability Engineering as early as possible, optimally with the first idea for a new medical product. Usability Engineering is not just a piece of a puzzle that fits in a certain place, it is a process that runs in parallel with several other processes like the design-, the quality-, or the risk-management process.

At the very beginning of a product development process, the manufacturer needs to get a good understanding of the different users, the use environment and how comparable devices are commonly used on a daily basis.

A well-established and easy to follow methodology to use during the first step is a contextual inquiry. A *contextual inquiry* is an interview technique that is conducted as a field observation in the actual users' workplace, like hospitals or, if no predecessor device exists, in other environments where similar devices are in use. The idea is to observe who the typical users are and how they use the device in the typical use environment. Follow-up interviews with the users can shed light on what difficulties and frustrations they face with these devices as well as any possible deviations from the manufacturers' intentions in their use.

During observation, the different groups of users should have become clear and members from each group should be interviewed so that *user profiles* can then be developed for each. These profiles should take into consideration demographic factors like gender, age, occupation and level of experience. Especially when patients are a user group in addition to health care professionals, it is important to consider certain impairments that could influence the use of the device. For example, Parkinson's patients may struggle to inject themselves due to tremors, muscular rigidity or slow, imprecise movements and diabetes patients may have a poor sense of touch due to a peripheral neuropathy.

In addition to allowing the creation of intended user profiles, observation should lead to the specification of typical *use environments*. Where and under what conditions is the device generally used? Any possible influencing factors, like the social context, lighting conditions, noise level, clinical versus home care use, and other attributes that may be relevant, should be included. Often the same device is used in different environments, like an infusion pump that is used in the intensive care unit, the operating room but also outside the hospital in an ambulance car, a helicopter, or a home care setting which all have specific, sometimes unique, influencing factors to the usage of such a pump.

If we think about an intensive care unit (ICU), we expect the light to be sufficient to read a displayed value and the hands of the user to be unencumbered. However, what if the infusion pump is being used in a helicopter, then we need to consider that sometimes, e.g., during take-off and landing no lights are allowed to switch on. If no internal display light is available, crucial parameters might not be readable. Or consider the use of the pump at the scene of an accident in the dead of winter: Paramedics may be wearing gloves which will make small buttons or touch screens difficult to use.

In addition to the definition of the intended user profiles and the use environment, the use specification is also the place to specify the patient population and the part of the body the device interacts with. Is the patient in a good condition, is the product intended to support kids or adults or both, or does the patient suffer from any allergy or intolerance. Indeed, there is also a difference if the device interacts intramuscularly, intravenously, or noninvasively.

If possible, it is helpful to already specify the general operating principle of the device in this phase. Are there mechanical or electronic components or both? Is radiation emitted? The manufacturer should also define the intended medical indication of the device as clearly as possible. Is the device intended to monitor, treat, diagnose, screen, or prevent a certain disease? What kind of disease or condition are we talking about?

Of course, during this first phase of the usability engineering process the users should get the chance to state their needs for a new product based on their experience and personnel preference. The results of the contextual inquiry must be evaluated and summarized. The user needs have to be transferred into product requirements and specifications for the product development (see "Establishment of the User Interface Specification" section). This information has to be documented for regulatory purposes but also helps the development team to understand what the intended user actually needs and why.

All attributes discussed above, the user profiles, the use environment, the patient population, the part of the body or tissue applied to or interacted with, the medical indication, and the operating principle, together define the medical device *use specification* according to the IEC 62366-1:2015. These are fundamental design inputs for identifying use errors as potential causes for hazards and hazardous situations related to the user interface and how they finally can contribute to harm (see next chapter).

Some of the information required at this stage may be a subset of the intended use description already specified in other documentation.

Identification of Hazards and Hazardous Situation

Once this initial investigation is complete, the next step in the user-centered design and development process is to evaluate possible hazards and hazardous situations when using the device in the intended way. This is similar to the procedure defined in the risk management standard for medical devices, ISO 14971. While ISO 14971 focuses on hazards due to technical issues, the evaluation required by IEC 62366-1:2015 focuses especially on hazards and hazardous situations resulting from use errors. The goal is to identify and describe the potential effect that a use error might have and how it can contribute to harm. A *hazardous situation* occurs if a person is exposed to this hazard. Use errors can directly lead to hazardous situations as a result of a user action or lack of action or indirectly as a result of the response of the medical device to the use error (see Figure 12.3). However, use errors do not always have to cause a hazardous situation or lead to harm.

Initially, characteristics of the user interface that can affect the safe use of the system must be identified. For medical products important considerations may include, but are not limited to, a definition of whether the system will be used in sterile or nonsterile environment, if the device is interpretative, if it includes software, if it includes mechanical and/or electrical parts, if radiation is emitted, and whether or not it is invasive. All this information can be found in the already conducted use specification.

Valuable methods that can be used at this point are *function analysis* and *task analysis*. The purpose of a function analysis is to identify those functions that are typically performed by the medical device itself and those that are assigned to the user. Some functions might require a shared responsibility by the device and the user. The function analysis should

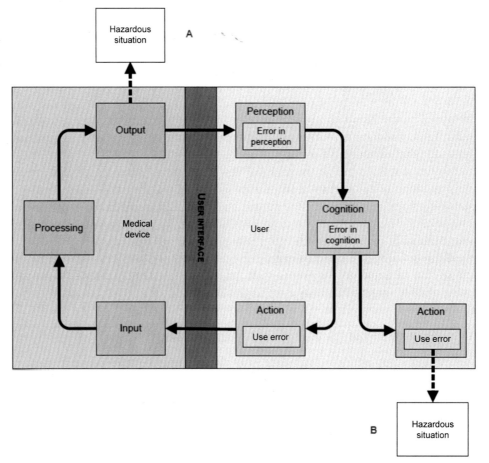

FIGURE 12.3 Model of user–medical device interaction (IEC, 2015): (A) hazardous situation caused by a response of this medical device; (B) hazardous situation caused by user action or lack of action on the patient or with a different medical device based on information obtained from this medical device.

consider at least all of the device's *primary operating functions* which means all those functions that are either frequently used or related to safety.

Those functions assigned to the user are called *tasks*. Fittingly, they are further discussed in the task analysis. The purpose of a task analysis is to identify a sequence of tasks the user needs to follow to complete an intended function. It furthermore defines the mental and physical activities the user needs to perform to complete a certain task, e.g., remember an ID, mentally calculate an infusion rate, press a button, or connect a sensor.

The results of the function and task analysis together with the results of the contextual inquiry (see "Definition of the Use Specification" section) help identify potential use errors that may occur. Of special interest for the identification of use errors is information about the different user profiles, e.g., if a device is used by healthcare professionals or lay users, and the context of use.

Besides the identified use errors, the investigation of hazards and hazardous situations should be based on information on hazards and hazardous situations that are foreseeable or that are known for existing medical devices with a user interface of a similar type and the identified use errors.

Description of Hazard-Related Use Scenarios

A further step in the usability engineering process is the definition and description of hazard-related use scenarios for the further evaluation steps. Generally, *use scenarios* describe a sequence of tasks performed by a specific user in a specific use environment as well as the device's expected responses. A hazard-related use scenario is a use scenario that directly or indirectly leads to a hazardous situation.

Figure 12.4 shows a use scenario leading to the expected response and the result intended by the user.

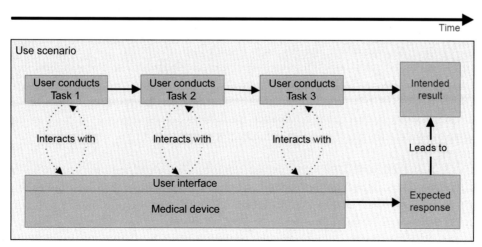

FIGURE 12.4 Use scenario.

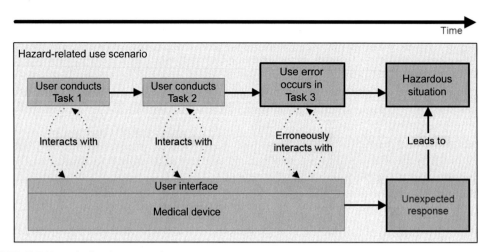

FIGURE 12.5 Hazard-related use scenario.

Figure 12.5 describes a hazard-related use scenario in which a use error occurs that leads to an unexpected response by the device that contributes to a hazardous situation.

The descriptions of the hazard-related use scenarios need to be based on the typical tasks the user will perform with the device, but must also include tasks that, while not intended by the manufacturer, are reasonably foreseeable.

For further evaluation, it can be beneficial to estimate the severity of the harm associated with the hazardous situation caused by the use error and the unexpected response of the device. This offers the opportunity to determine a subset of hazard-related use scenarios for the summative evaluation which is based on the severity of the potential harm.

Finally, as a measure of the risk management process, the manufacturer of a medical device has to investigate how potential use errors resulting in harm can be mitigated effectively. The best way to mitigate the risk is to ensure that the design does not allow for any use error to occur. An early integration of user-centered design activities allows these considerations to be realized easily. Is an error preventative design of the medical device not possible, the manufacturer has to establish protective measures in the device itself or in the manufacturing process. If there is no chance for further design improvements or protective measures, the final option is to offer information for safety, e.g., in the user manual or on the device itself.

Another real life example: Five cases are known where children were hooked up to cardio-respiratory monitors in their own homes were severely injured or died when the device's electrodes were plugged into normal sockets. The home's alternating current resulted in deadly arrhythmias. Only after these tragedies had occurred did the manufacturer replace the open electrodes with plastic-coated safety plugs as shown in Figure 12.6. Risk mitigation needs to happen early on so that such tragedies can be avoided.

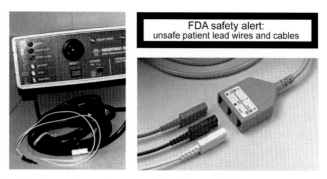

FIGURE 12.6 Electrode cable of a cardio-respiratory monitor before and after a mortal incident.

Establishment of the User Interface Specification

As described in "Description of Hazard-Related Use Scenarios" section, it is important to take potential hazards and hazardous situations into consideration as early possible, so that risk-mitigation measures may be taken into consideration in the design process from the start. These measures are especially relevant for the specification of the user interface. Risk mitigations can directly contribute to the user interface specification defining, for example, the coding of connectors in color or shape, the necessity of training before using a product the first time or the use of an access code to modify safety relevant functions. Other *user interface specifications* can include, but are not limited to, the definition of a font size on graphical user interfaces, the size of certain buttons, the color-coding of graphical elements, the use of symbols and labeling or the dimensions of components or the device itself. Font types, color schemes, or the structure and shape of buttons are often defined in a design style guide. Further usability relevant specifications can be found, e.g., in the design or product requirement specification.

The law requires manufacturers to produce a safe design, it does not, however, speak to ease, joy or effectiveness of use which are typical parameters of usability engineering in other domains like software engineering or the automotive industry.

Of course safety should be the primary consideration, but an investment in user-friendliness is sure to pay off for manufacturers as well. Few devices are unique in function anymore, and users who now have options have gotten critical in recent years; they want devices that meet their needs and make their daily work easier. Ease of use can help manufacturers not only produce a device that they may legally sell, but much more one that people want to use—and buy.

When talking about user interface specifications, it is important to transform the user needs, design requirements, and risk mitigations into measures that are evaluable. It is not sufficient to ask for a device that is easy and intuitive to use. These statements need to be formulated as clear specifications. What makes a device easy to use? Does the number of steps to activate a function have to be as small as possible? Do all buttons need to be clearly labeled with text instead of symbols? Does the system have to have a haptic coding for all connectors? Is it necessary that the system not require tools to exchange certain components? Does the system have to be operable under dimmed lighting conditions?

In the user interface specification, a manufacturer needs to define all parameters that make a device safe and easy to use, e.g., the user must be able to activate the defibrillator within 30 seconds under various environment conditions, e.g., in the OR, on the ward, in the ambulance car, or outside in any other environment. The climate, the noise level, and the lighting conditions might need to be specified too. If we're talking about an AED that should be usable by just about everybody, it is especially important to also define the user groups and special needs that that entails.

As the user profile, the use environment as well as the patient are important aspects that need to be considered for the user interface specification, the previously defined use specification, containing a definition of all these attributes and the use scenarios, is of great importance. As previously explicated, one objective is to reduce the likelihood of a use error associated with the medical device to a minimum.

Often the instructions for use (IFU) or a training are included as a risk control measures. In this case, or generally, if an IFU will be provided by the manufacturer and a training is mandatory, this needs to be addressed in the user interface specification too.

The user interface specification required by IEC 62366-1:2015 may be integrated into other specifications.

Within the following design process, the user interface specification might be subject to change based on the results of the formative evaluations with potential users.

Performance of Iterative Usability Evaluation

One of the most important principles of usability engineering is the integration of typical or potential user groups early in the product design and development process—hence the synonymous reference to user-centered design. Instead of a decision-making process within the project team, it is much more effective to ask users about their opinion and their preferences.

Formative Usability Evaluation

User involvement is possible very early on, with conceptual models or even first scribbles of a new product concept. Focus group discussions are one of several methods to get feedback from a couple of people who meet the criteria of potential users. It does not matter if they are naïve users or professionals, if they are patients, clinicians, or technicians, male or female, as long as they are specified as potential users in the use specification. The goal of these early evaluations is to select the preferred design among several competing concepts based on the user needs, collecting suggestions for improvement and defining the right path for the further product design.

These kinds of evaluations at an early stage of the development process serve as guideposts toward a successful device. The feedback can be used to double-check the user needs and the user interface specification but might also point out new potential use-errors or hazardous situations and required design modifications to mitigate the associated risk.

The iterative nature of the usability engineering process now requires that the product concept and the user interface specifications as well as the risk analysis will be updated based on the evaluation results. This might lead to an updated design, made available as a mockup or software simulation, that still offers room for improvements in another formative evaluation.

The next round of user involvement might use semi-structured interview techniques, including a hands-on session that gives the respondents the opportunity to explore the new concept and check if it meets their individual requirements or if additional improvements are necessary. During the informal evaluation session, the moderator can use open and closed questions to get additional feedback from the interviewees.

It is beneficial to develop a formative usability evaluation plan that defines the objective and the scope of the study. Furthermore, the involvement of users should be specified and a description of the evaluation method as well as a list of questions and important aspects that need to be discussed should be included.

The number of formative evaluations and even the number of users involved in the individual studies can vary depending on the complexity of the device, the number of user profiles, or the context of use.

Formative usability evaluation sessions can focus only on the hardware or software, represented as early scribbles or wireframes, form factor models or click dummies, look alike models or software simulations. They can also strictly consider individual components of the system, accessories or the instructions for use or the training material. The question is always "Is the design appropriate or are there any improvements that make the product safer, more effective or easier to use?"

Naturally, the manufacturer sometimes has to judge if all suggestions for improvement can be considered or if a compromise needs to be found, e.g., because of a conflict of interest between different user groups, commercial reasons, or deviations from regulatory requirements.

The probability of a conflict of interest is high if the product needs to fit several user groups or use environments. Consider an OR navigation system that needs to be used by orthopedists, neurologists, and spine surgeons or an infusion pump that allows for use in the ICU, neonatal care but also in the OR and rescue services. Intercultural differences are an additional important variable to consider. Manufacturers eager to distribute their product worldwide need to consider that the education and qualification in the different distribution countries may vary. Ventilators used by intensive care nurses in Germany are controlled by specially educated respiratory therapists in the United States. The same goes for the complete public health and insurance system affecting the processes and workflow in a hospital or other institutions of the public or private health system.

Cultural differences can also be found in the meaning of colors, the interpretation of symbols, and cognitive processes and expectations. Take the color red: In the Western Culture, we see it as a clear warning color, but not everyone will agree. In China and other Asian

 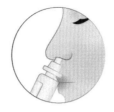 

FIGURE 12.7 Different meaning of graphical instructions depending on the reading direction.

countries, it is a symbol of happiness. This means instead of the "Stop" button, in Asian countries the "Start" button might be red.

Another example is miniaturization: Say your company has produced a compact, high-quality device that is well-received on the European market, you might expect that it will also sell well in Asian countries. However, this assumption does not take into account a slightly the different mind-set of the culture where a product is always expected to bring "good value for money." You would be better off not selling a miniaturized version of your device in the Far East but instead you need to increase the product's weight and volume.

Some languages are also read from right to left rather than left to right, as is common in the Western culture. In the following example (see Figure 12.7), the manufacturer tried to design an international instruction for use of a nasal spray without any textual information. But reading the graphical information from the right to the left (e.g., as in most Arabic-speaking countries) might imply to the reader that the spray should be used in case of a dry mucous membrane of the nose.

Summative Usability Evaluation

Once the design is final and no further formative evaluation sessions are planned, the final product needs to be evaluated to determine if it is safe to use as intended. This last step of the usability engineering process is called *summative usability evaluation*.

At this stage, it is important to have the final product available, including the intended packaging and labeling, all required accessories and the accompanying documents. The production is based on the final tools and the embedded software is in its final version. The labeling and packaging, e.g., of an automatic injection device, is final. Safety labels and symbols are incorporated. The IFU is available in its final version, either printed or electronically, if e-labeling is allowed. If a training is mandatory, the training script to be used is available.

To judge if the system can be used without any unacceptable risk, it is important to simulate a realistic environment taking into consideration all possible factors that might influence the safe use of the device. This can include, but is not limited to, the noise level, the lighting conditions, the social environment, the spatial conditions, other devices that interact with the system under test or typical clothes that might influence, e.g., tactile feedback or movements. In addition, training needs to be provided in advance of the test session if this mirrors the real world situation.

All intended user profiles need to be considered during the summative evaluation. Furthermore, all hazard-related use scenarios or a selection based on the severity of the associated harm have to be considered and it needs to be evaluated if the determined hazards and risks have been mitigated successfully. This also includes the verification of the effectiveness of the IFU or a training if those are indicated in the risk management file as risk mitigations.

In case additional risks can be observed while the test participants are interacting with the device, the risk management file must be updated and an analysis of the new issues has to be made.

A method described in various publications and standards is the *root cause analysis.* During the one-on-one test session, the moderator has to observe if the test participant is using the system correctly. If the moderator observes a use error, the test participant will be informed and the observed issue will be discussed in detail. The root cause of the use error is determined based on the observations of user performance by usability experts and the subjective comments of the test participants. If new hazards, hazardous situations or hazard-related use scenarios result, the manufacturer has to determine if further improvement of the user interface as it relates to safety is necessary or practicable. If yes, the device needs to reenter the usability engineering process; if not the manufacturer has to provide sufficient evidence that the residual risk is acceptable.

Optionally, but not mandatorily, the manufacturer can choose to define objective acceptance criteria, e.g., "90% of the respondents were able to inject the intended dose" or subjective criteria like "80% of the test participants rate the ease of use a 2 on a scale of 1−5, where 1 is the best."

Beyond any quantitative measures, be they subjective or objective, the major success criterion is use without any unacceptable risk.

The summative usability evaluation needs to be organized and documented by developing a summative usability evaluation plan. This plan must describe at least the evaluation method, the part of the user interface that will be evaluated, the involvement of the intended user profiles, the conditions of use, the availability of an IFU, and a provision of training in advance of the summative evaluation and a description of the selected hazard-related use scenarios.

The results of the summative usability evaluation, including the root cause analysis, will be documented in a summative usability evaluation report.

The Usability Engineering File

As required by the usability standard IEC 62366, the manufacturer has to document the usability engineering process. This documentation is called the *usability engineering file.* It has to contain certain statements regarding the process in general (see IEC 62366, Chapter 4) as well as a documentation of every single step of the usability engineering process (see IEC 62366, Chapter 5). The usability engineering file does not have to physically contain all the information. More often you can set a link to other documentation of your product development process. For regulatory purposes, it is recommended that the link is obvious and clearly presents the required information. Furthermore, it is crucial to provide traceability over all documents especially between the use-related hazards and risks and the summative usability evaluation.

The easiest way to provide good usability documentation that will lead to a successful certification is to understand that usability engineering is not one piece of a big puzzle. Usability engineering is a process that starts early in the development process. A manufacturer has to engage heart and soul in usability engineering, although parts of the process are not completely new. Some of the required process steps and requests for documentation are already issued in other standards.

A final piece of advice—the sooner you fail, the more quickly you'll reach your goal.

Abbreviations

AED Automated External Defibrillator
AOK Allgemeine Ortskrankenkasse
CIRS Critical Incident Reporting System
FDA Food and Drug Administration
HFE Human Factors Engineering
ICU Intensive Care Unit
IEC International Electrotechnical Commission
IFU Instructions for Use
IOM Institute of Medicine
MDD Medical Device Directive (Medizinprodukterichtlinie)
OR Operating Room

References

Institute of Medicine, 2000. , 0-309-06837-1. To Err is Human - Building a Safer Health System. National Academic Press.

AOK Krankenhaus-Report 2014: "Wege zu mehr Patientensicherheit."

<http://www.sanders.senate.gov/newsroom/video-audio/medical-mistakes-are-3rd-leading-cause-of-death> (retrieved October 28, 2015).

James, J.T., 2013. A new, evidence-based estimate of patient harms associated with hospital care. J. Patient Sat. 9 (3), 122–128.

IEC 62366-1:2015 "Medical devices—Part 1: Application of usability engineering to medical devices." Annex A (informative); IEC Central Office, published February 25, 2015.

13

Clinical Engineering Education and Careers: Overview at the University of Connecticut

Frank R. Painter

CLINICAL ENGINEERING, UNIVERSITY OF CONNECTICUT, STORRS, CT, USA

Introduction

The American College of Clinical Engineering (ACCE, 1991) defines clinical engineering as follows: *"A Clinical Engineer is a professional who supports and advances patient care by applying engineering and managerial skills to healthcare technology."*

The education of clinical engineers has been primarily accomplished at the graduate level. Purdue University and Case Western Reserve had outstanding graduate educational programs in clinical engineering for many years. The principal professors in these programs retired and the programs faded. Currently, there are active graduate clinical engineering programs at the University of Toronto, Canada, the University of Trieste, Italy, and the University of Connecticut, Storrs, CT, USA. These programs have an academic- and internship-based component and graduate 5 to 10 students per year each. There may be other programs with some clinical engineering content, but they primarily focus on biomedical engineering, product design, or manufacturing support.

The need for clinical engineers is clear to those who understand the role of technology in healthcare and the need to support this technology throughout its life-cycle. Having clinical engineers available to become involved in technology assessment, acquisition, implementation, on-going management, reducing the risk of ownership, identifying appropriate replacement strategies, and insuring users are supported in their use of technology is sometimes overlooked. These roles may sometimes be left to other individuals in the health care organization (HCO), to vendors who provide the technology, or it may be that no organized management is provided except that provided by individual users. In most cases, HCOs without clinical engineers have little or no centralized management of these important functions.

Clinical Engineering. DOI: http://dx.doi.org/10.1016/B978-0-12-803767-6.00013-1

An ideal clinical engineer should be an expert in project management, a technical expert in the types of technology owned by the HCO, someone who understands "systems thinking" and who applies systems engineering principles to solve problems in the health care environment, a health care technology manager who has a firm grasp on the principles of risk management, financial management, personal management, vendor management, service management, and regulations management. The clinical engineer should also have a background which enables them to understand the principle challenges faced in the health care work environment such as work flow, infection control, medication errors, staffing issues, capital versus operating expenses, cash flow constraints, and local, national, and international regulatory constraints. Creating an educational program to graduate engineers who can provide a meaningful contribution to the health care environment in these areas is a challenge, but one which can be achieved with institutional and regional professional support.

As can be imagined, the approach to educating young engineers in each of the areas mentioned can be somewhat variable. Ideally the curriculum should be designed to match the needed skill set a practicing clinical engineer will require to function on the job. Conveniently, the American College of Clinical Engineering needed to quantify the clinical engineering "body of knowledge" (BOK) in their support of the clinical engineering certification program they help administrate. They did this by sending a survey to over 500 clinical engineers worldwide asking what skills were important in their jobs and what portion of their jobs were spent in each area of clinical engineering responsibilities. Over 350 responses resulted in a statistically significant result. This process has been repeated every 3—4 years over the past 15 years, resulting in a solid set of data which can be used to design an educational program and track changes in the profession as they occur, so educational programs can adjust as needed (Cohen, 2002).

From this data (see Box 13.1—*Body of Knowledge*), the UCONN program has created the current graduate clinical engineering curriculum and from the changes in the BOK seen over the years, new courses have been added or existing courses have been modified. International clinical engineers have responded to the survey and their data has been included when reporting the current state of the body of knowledge, but a more focused survey including data from just a single region of interest could result in a useful data set for just Europe or Latin America or the southeast Asia region, for instance. Some variation in the BOK from region to region should be expected because of varying financial resources or specialized codes and standards, but the variations would be relatively minor in the overall scheme of things. As time goes forward and we see the world community's efforts to harmonize standards and practices globally gaining a foothold, the differences in the BOK will be further reduced.

Another consideration when designing a clinical engineering educational program is deciding to what level we educate clinical engineers before sending them off into the workplace. Two considerations which come to mind are first, the educational programs at the

Chapter 13 • Clinical Engineering Education and Careers   185## BOX 13.1 BODY OF KNOWLEDGE

1. **Technology Management—32%**
   a. Product Selection/Vendor Selection
   b. Technology Assessment
   c. Project Management
   d. Capital Planning
   e. Interpretation of Codes and Standards
   f. Usability/Compatibility Assessment
   g. Healthcare Technology Strategic Planning
   h. Clinical Device Use and/or Application
   i. Device/System Upgrade Planning
   j. Device Integration Planning
   k. Clinical Systems Networking
   l. Life Cycle Analysis
   m. Coordinating Device Interoperability/Interfacing
   n. Other Technology Management Responsibilities
   o. Return on Investment (ROI) Analysis
   p. EMI/RFI Management
   q. Pre-clinical Procedure Set-up/Testing
   r. Clinical Trials Management (Non-investigational)
   s. Water Quality Management
   t. Participation in Clinical Procedures (e.g., surgery)
2. **Service Delivery Management—17%**
   a. Technician/Service Supervision
   b. Equipment Repair and Maintenance
   c. Equipment Acceptance
   d. Service Contract Management
   e. Equipment Performance Testing
   f. Maintenance Software (CMMS) Administration
   g. Develop Test/Calibration/Maintenance Procedures
   h. Parts/Supplies Purchase and/or Inventory Management
   i. Other Service Delivery Responsibilities
   j. Technical Library/Service Manuals Management
3. **Product Development, Testing, Evaluation, and Regulatory Compliance—5%**
   a. Regulatory Compliance Activities
   b. New Product Testing and Evaluation
   c. Documentation Development/Management
   d. Human Factors Engineering
   e. Product/Systems Quality Management
   f. Device Modifications
   g. Medical Device Design
   h. Product Research and Development
   i. Medical Device Concept Development/Invention

      j. Other Product Development Responsibilities

      k. Product Sales/Sales Support

4. IT/Telecom—8%

      a. Integration of Medical Device Data

      b. Information Technology (IT) Management

      c. Help Desk/Dispatching/Call Tracking

      d. Other IT/Telecommunications Responsibilities

      e. Telecommunications Management

5. Education of Others—10%

      a. Technician Education

      b. Device User/Nurse Training

      c. Develop/Manage Staff Training Plan

      d. Engineering Education

      e. Other Education Responsibilities

      f. International Health care Technology Management

6. Facilities Management—5%

      a. Facility Emergency Preparedness Activities

      b. Emergency Electrical Power

      c. Building Plan Review

      d. Medical Gas System Testing

      e. Building Design

      f. Other Facility Management Responsibilities

      g. Facility/Utility Remediation Planning

      h. Supervise/Manage/Direct Facilities Management

7. Risk Management/Safety—11%

      a. Patient Safety

      b. Product Safety/Hazard Alerts/Recalls

      c. Incident/Untoward Event Investigation

      d. Engineering Assessment of Medical Device Failures

      e. Risk Management

      f. Root Cause Analysis

      g. Medical Device Incident Reporting (SMDA)

      h. Infection Control Failure Mode and Effect Analysis

      i. Workplace Safety Practices (OSHA)

      j. Fire Protection/Safety (Life Safety Code)

      k. Radiation Safety

      l. Hazardous Materials

      m. Industrial Hygiene

      n. Other Risk Management/Safety Responsibilities

      o. Expert Witness

      p. Investigational Research (Human Use)

      q. Forensic Investigations

8. General Management—11%
   a. Budget Development/Execution
   b. Personnel Management/Supervision
   c. Staffing
   d. Staff Skills/Competency Assessment
   e. Policy/Procedure Management/Development
   f. Performance Improvement/CQI
   g. Business/Operation Plan Development/Management
   h. Committee Management
   i. Other General Management Activities
   j. Revenue Producing Activities
9. Other—1%

From *ACCE BOK Survey 2011.*

undergraduate level primarily focus on educating young engineers in the classical areas of biomedical, electrical, mechanical, or some other engineering field. These programs provide a broad but substantial education in fundamental areas an individual would need to graduate in that engineering specialty. The time available to learn about peripheral areas of study (e.g., clinical engineering) would be limited, but at the undergraduate level a series of clinical engineering orientation classes or a course in clinical engineering could nicely fit in to a classical engineering undergraduate degree. The time of specialization and further academic preparation is at the graduate level. Since there is so much to learn to function well as an engineer in the clinical environment, to properly prepare a clinical engineer requires a graduate engineering education.

The other consideration when preparing clinical engineers is the professional environment of the health care system. Most of the practitioners in the health care management environment, including physicians, administrators, nurse managers, financial managers, and clinical managers receive education at the graduate level. In order for clinical engineering to integrate well at this level of the organization, a similar degree of educational advancement needs to be achieved. Generally speaking, when clinical engineers come into the organization with an MS degree, they have a much stronger level of influence in and acceptance by the organization.

The overall clinical engineering program design is based on a balance between academic course work and practical experience by way of an internship. By taking academic courses and participating in an internship program, the student will benefit from two primary mentors and teachers, firstly the UCONN Clinical Engineering Program director who teaches most of their academic courses, but also the hospital-based clinical engineering department director who oversees their daily work activities. Learning the information in the class room setting and applying it in the health care setting gives the student a firmer grasp on the material. As a result, upon graduation, the students who enter the hospital environment are well prepared and rarely encounter a situation or activity with which they are not already familiar. After this level of preparation, they are ready to "hit the ground running" so to speak.

Curriculum

The curriculum has been designed to provide the skills and information needed for a clinical engineer to function on the job in a HCO. The general guideline used to guide curriculum definition is the ACCE Clinical Engineering BOK survey results (Cohen, 2002). There are seven clinical engineering graduate level courses delivered over a four semester program at UCONN which cover nearly every aspect of the BOK. There are several themes which run throughout the program, including engineering management, systems engineering, understanding health care delivery systems, the role of regulators, the role of standards making organizations, the role of professional organizations, the role of manufacturers and suppliers, and the function and use of each primary technology in the HC environment. Five of these seven courses include 12 three-hour classes, homework, a semester project which requires a presentation and paper written on a topic related to the course material, a midterm exam, and final exam. They are as follows:

Clinical Engineering Fundamentals
   Provides the fundamental concepts involved in managing medical technology, establishing and operating a clinical engineering department, and the role of the clinical engineering department in supporting the clinician in the use of technology. Topics covered include:
- Technology management
- Technology assessment, acquisition, and implementation
- Service management
- Regulations, codes, and standards
- Budgeting and financial management
- Safety
- Risk management
- Performance improvement
- Personnel management
- Equipment replacement planning
- Computerized maintenance management systems
- Medical equipment planning for new facilities.
Engineering Problems in the Hospital

This course covers engineering solutions to problems that are found in the health care environment and provides a basic understanding of utility systems which directly connect to or profoundly impact medical equipment. This includes a wide variety of topics such as:

- Ventilation and indoor air quality
- Infection control
- Medical gas distribution systems
- Emergency management
- Electrical power quality and electrical supply systems in hospitals
- Uninterruptable power supplies
- Electrical safety in the patient care environment
- Project management
- Electromagnetic interference and electromagnetic compatibility
- Radiation shielding and radiation protection
- Lighting
- Real-time location systems
- Hospital fire protection systems
- Telemedicine and medical image transmission
- Hospital architecture and the design of patient care facilities.

Human Error and Medical Device Accidents

This course teaches the basic principles needed to analyze medical devices, medical device users, medical device environments involved in medical device accidents. It particularly focuses on human factors engineering as an important step to minimizing human error and the role of systems engineering thinking in reducing medical device use errors and medical device accidents. Topics to be covered include:

- Patient safety
- Device development
- The role of the FDA
- Human factors engineering
- Types of human error and taxonomy of medical device accidents
- Root cause analysis
- Failure modes and effects analysis
- Accident and incident investigation
- Legal aspects of medical device-related injuries
- Operating room fires
- Electrosurgical burns, laser burns, and tissue injury in the medical environment
- Anesthesia injuries
- Infusion device accidents
- Catheters and electrode failures.

Medical Instrumentation in the Hospital

This course will examine 8—10 current major technologies in use by health care practitioners. It will review the physiological principles behind each technology, the principles of operation, major features, methods for testing and evaluating each technology, and highlight key considerations when managing these technologies. Some classes involve site visits to observe and examine the equipment being discussed. Technologies to be covered will be selected from:

- Anesthesia equipment
- Medical imaging fundamentals
- Radiographic and fluoroscopic equipment
- Magnetic resonance imaging
- Nuclear medicine gamma cameras
- Computed tomography and PET/CT
- Linear accelerators
- Mammography
- Ultrasound imaging equipment
- Laboratory instrumentation
- Surgical and ophthalmic lasers
- Dialysis equipment
- Cardiac assist devices
- Surgical and endoscopic video systems.

Clinical Systems Engineering

Primarily covers medical device connectivity and interoperability. This includes connecting medical devices to the hospital network to pass data to the patient medical record and connecting one medical device or system to another for the purpose of feedback and control. Subjects to be covered include:

- Basic networking concepts
- Medical systems security and risk management
- HL7 and DICOM standards
- Middleware
- Connectivity standards and methods
- Medical device interoperability
- Medical device integration
- Clinical information systems
- Digital imaging and image storage systems
- Medical device integration project walk-through.

Clinical Rotations I and II

This course follows a different format. It includes 24 weeks of 4—5 hour observations in the clinical environment. It consists of clinical engineering rotations that the interns arrange in hospitals as they observe the technology/patient/clinician interface in areas such as:

- Surgery/operating room
- Anesthesiology

- ICU
- Diagnostic imaging
- Laboratory
- Gastroenterology
- Ophthalmology laser clinic
- Radiation medicine
- Electrophysiology
- Emergency room
- Clinics
- Homecare
- Dialysis
- Electrocardiography
- Physical therapy
- General medical and surgical floors.

Clinical Engineering Internship

Although the academic portion of the program is important in providing information to prepare young engineers for the field of clinical engineering, the real value in the program lies in being able to participate in the internship where the information received is applied in the actual health care setting. During the internship, the student is mentored by the director of clinical engineering at the hospital where they are located during the 20-month period from the beginning of the first semester to the end of the fourth semester. After their initial period of orientation, the students are considered to be clinical engineering department staff and are given responsibility and accountability. This experience helps them apply the information they learned in class and acquire other skills and knowledge to accomplish the assigned responsibilities.

There are 14 hospitals which take clinical engineering interns. Some take two interns per year, some take one new intern every year, and some take an intern every 2 years. The arrangement between the university and the hospital is contractual. The hospitals pay the university for each student working in the hospital. The university in turn pays the student a graduate student stipend and waives their tuition. This pay arrangement allows the student to then be treated as a paid employee with the corresponding obligations and accountability.

Around 40–50 students apply to the program each year. There are typically 10 to 12 funded positions, so we accept about 15 candidates to interview for the openings. These internship positions are treated as job openings so each hospital clinical engineering director with an opening interviews all the candidates. At the end of the interview process, which usually takes three to four weeks to complete, each candidate rates the hospitals and each hospital rates the candidates. After a matching process, the students who are accepted for an internship position are informed of the outcome and assigned to the hospital that chose them.

As any other employee in the hospital, the student must take and pass a pre-employment physical and take a series of orientation classes to explain basic operational issues in the hospital such as fire safety, infection control, physical security, computer security, general safety, employee policies, and hospital organizational structure.

Within a short period of time, the student is considered to be a department staff clinical engineer and given responsibilities with equipment acquisition and implementation projects, with administrative duties such as managing the CMMS, the equipment replacement planning program, or other data-related projects and in attending committee meetings, occasionally representing the department at these meetings. An example of many of the possible work assignments is listed in Box 13.2.

BOX 13.2 CLINICAL ENGINEERING WORK ASSIGNMENTS IN THE HOSPITAL

It is expected that the intern will be assigned to participate in the majority of the following activities at some point in their 2-year program.

- Establish a basic understanding of general medical equipment through 2–3 months of shadowing BMETs, performing inspections and minor repairs of a variety of devices contained in checklist of basic medical devices
- Develop new equipment inspection procedures
- Review and update/expand (if appropriate) department policy and procedures manual.
- Review/update employee job descriptions (if appropriate)
- Participate in an employee evaluation process (if appropriate—with consent)
- Prepare at least two short CE department staff in-service presentations (one per year) to teach
- Become familiar with JCAHO technology management standards and compare and comment on department practices designed to meet the standards
- Participate in department based JCAHO mock survey and participate in resolution of problems found
- Participate in risk assessment of new technology for JCAHO inclusion
- Participate in the department's competency assurance program
- Participate in at least one HFMEA or RCA development process
- Participate in (and eventually lead if appropriate) department performance improvement program data collection process, including among other things a customer satisfaction survey
- Accompany department director to hospital safety committee; technology selection committee; capital planning committee; and other committee meetings as appropriate
- Accompany department director to one department management meeting, hospital management meeting and department director's one-on-one meeting with their administrator
- Make one presentation on behalf of the department to higher level hospital managers
- Be given the opportunity to interact with outside agencies, vendors, or consultants
- Technology assessment to evaluate appropriateness of device to meet clinical need
- Evaluation of equipment for purchase including life-cycle cost analysis report, total cost of ownership report or new technology business plan

- Incoming inspections of new equipment or systems
- Installation of new equipment or systems (or oversee installation)
- Clinical staff in-service training program development (or oversee vendor training)
- Participate in the hospital's equipment replacement planning process
- Participate in the process to manage the CE department's website
- Participate in the development/management of the CE department's computerized medical equipment management system
- Participate in a hospital expansion/renovation project, becoming involved review of the architectural, engineering and equipment selection parts of it
- Participate (if appropriate) in development of the annual department budget.
- Review codes and standards to evaluate the department's/hospital's regulatory compliance
- Participate in the management of an extended project
- Participate in the management of recalls and alerts program
- Participate in the evaluation of several service contracts
- Participate in the investigation of at least one incident involving a medical device.

Although the student may be one of the newer employees in the department, after a period of time they may be asked to lead special technology projects, using the resources of the department to complete the work. This requires them to develop the ability to organize the work to be done, work with limited resources, communicate well with all levels of the organization, and use others to accomplish parts of the work.

The internship is organized to take advantage of being located in the hospital setting. The student is required to work a productive 20 hours per week for the department to which they are assigned. Sometimes productivity is reduced for one reason or another, so students usually work a few extra hours each week to ensure obligations are fulfilled. Additionally, each student is required to do about five hours of clinical rotations each week as part of their Clinical Rotations course requirement this time is outside their normal work responsibilities. The students are also asked to do all of their classwork, research, and writing for the academic portion of the program while they are in the hospital. This usually amounts to another 5−8 hours per week on average on-site making them available in the hospital should an important event occur from which they might benefit by being involved. So the end result is that students spend between 30−35 hours per week in the hospital clinical engineering department.

Employment

The majority of the second year clinical engineering internship program students receive multiple employment offers before graduation and most (∼75%) secure jobs before graduation. The practical experience received during the internship and the academic preparation focusing on the subjects clinical engineers need to know in the work place, make the graduates highly appealing to employers familiar with the program and well prepared to interview with employers who are not familiar with the program.

Preparation for the job search process begins about six months before graduation when each student fine tunes their resume and cover letters to potential employers. The students obtain business cards, practice their interviewing skills, and receive training on how to approach the job search process. Knowing that most jobs are obtained through networking with those in the field, students begin the process of meeting as many clinical engineering contacts as they can about a year in advance of graduation. An excellent way to make contact with other professionals in the clinical engineering profession is through professional societies. Students are required to join the regional clinical engineering society (New England Society of Clinical Engineering), the national professional society for clinical engineering (American College of Clinical Engineering), and the national trade organization for healthcare technology (Association for the Advancement of Medical Instrumentation). Meetings of these organizations provide good opportunities to meet potential employers or people who might know where the current job openings are. Additionally, these contacts can be the basis for ongoing professional relationships throughout an individual's career, an invaluable professional advantage.

The vast majority of UCONN's clinical engineering internship program's graduates initially get jobs in hospitals working in the clinical engineering department. A few have taken positions in the IT department integrating medical equipment, in the equipment planning department to assist in facility expansion, in the safety and risk management department managing technology risks, and in the imaging department managing the PACS system as well as acquisition and integration projects. Some have been employed by medical equipment manufacturers as clinical support specialists, in product service management, or in design and installation support for the sales team. A few have taken jobs with equipment planning companies with responsibilities in medical equipment planning, specifying and installation management for large construction projects. Two have taken positions with the government, in the FDA, as medical equipment specialists. The salaries of new graduates from this program compare favorably and sometimes exceed the average starting salary for engineering graduates with an MS degree in other fields.

Employment Market

An interesting observation is that many regions of the country do not have clinical engineers on staff in the hospitals at all, yet the hospitals in other regions utilize many clinical engineers and in fact could not function at the level they have been accustomed to without them. An example of this is that most of the central United States, outside of very large cities, has very few clinical engineers working in hospitals, if any. Yet in the northeastern United States and in most major metropolitan areas, most hospitals over 350 beds have a clinical engineer on staff. In the communities surrounding the University of Connecticut, the following examples tell the story of how having qualified candidates immediately available has resulted in great growth over the past 10−15 years.

Baystate Health Systems, a 716-bed hospital system in Western Massachusetts, initially had three clinical engineers in the year 2000. Now they have three on the medical device integration team, three in clinical engineering, and one in administration. Hartford Hospital, an 819-bed hospital in central Connecticut, had three clinical engineers on staff in the year 2000, one as director of clinical engineering, one as director of engineering, and one in the laboratory. Today they have five in the department of clinical engineering. Massachusetts General Hospital, a 950-bed medical center in eastern Massachusetts, has been involved with clinical engineering education for many years. There are seven clinical engineers working for the CE department now. In addition, the system to which MGH belongs, Partners Health Care, has six clinical engineers at their Brigham and Women's Hospital location, as well as four on the medical device integration team and three in the simulation center. Yale New Haven Hospital, a 1541-bed health care system in southern Connecticut, at one point in the recent past had no clinical engineers. Because of changes in the clinical engineering department leadership, they now have five clinical engineers on staff. This growth across the region is likely due to the fine track record of cost-effective use of clinical engineers to improve the quality of the clinical operations, resulting in hospital administration seeing the value. Another possible reason is that the CE directors who are the direct supervisors and mentors to the CE internship students recognize a well-qualified and trained team member in the student they have gotten to know for 2 years and can passionately recommend that the hospital hire them to continue progressing.

Final comments on the job market for clinical engineers in the United States are that the availability of properly trained and oriented clinical engineers is the limiting factor in the expansion of the profession. If there were more, they would be hired immediately. If the educational system could interest young people in the field, if colleges and universities were able to expand clinical engineering education at the BS and MS level and if clinical engineering department directors would offer their departments as a training ground during the educational process, the number of practicing clinical engineers could easily double. Putting these systems in place would accelerate the process.

Conclusion

Creating clinical systems engineers at the University of Connecticut has been a highly successful endeavour. Although it is the passion of only a few individuals at UCONN, it has been producing clinical engineers of the highest caliber. To be truly successful and better serve the entire health care community, clinical engineering education needs to spread to other major metropolitan areas where there are hospitals with functioning clinical engineering departments. The directors in these departments can not only take students, but act as partners with the university in sculpting curriculum and providing specialists to teach some of the material. This model can be duplicated. If that were to happen, the benefits to society and the health care industry would be immeasurable.

References

American College of Clinical Engineering. 1991 — http://accenet.org/about/Pages/ClinicalEngineer.aspx.

Cohen, T., 2002. ACCE Body of Knowledge Survey. J. Clin. Eng. Fall. 27 (4), 298—299.

The New England Society of Clinical Engineering (NESCE) — https://nesce.org/.

American College of Clinical Engineering (ACCE) — http://accenet.org/.

Association for the Advancement of Medical Instrumentation — http://www.aami.org/.

14

Certification of Clinical Engineers

Mario Medvedec[1], James O. Wear[2]

[1]UNIVERSITY HOSPITAL CENTRE ZAGREB, DEPARTMENT OF NUCLEAR MEDICINE AND RADIATION PROTECTION, ZAGREB, CROATIA [2]SCIENTIFIC ENTERPRISES, NORTH LITTLE ROCK, AR, USA

Introduction

Biomedical engineering integrates physical, mathematical, and life sciences with engineering sciences applicable in biology and medicine, for the purpose of preserving or improving health as a state of complete physical, mental, and social well-being and not merely the absence of disease or infirmity. Clinical engineering is the specialty within biomedical engineering, and it deals with all possible aspects of medical devices, equipment, systems, and overall health technologies used in hospitals or other clinical settings (Biomedea, 2013). Medical doctor Caesar Caceres actually coined the term "Clinical Engineer" in 1967 for various types of engineers, physical scientists, physiologists, and other professionals working with physicians in the clinical environment and performing medical and biological engineering type of work. Due to the scarce university study programs in clinical engineering leading to diploma or degree explicitly in clinical engineering, the term "Clinical Engineer" in this chapter normally refers to an occupation and profession rather than to an exact academic title or degree.

Clinical engineers are a subset of Biomedical Engineers who are, according to the International Standard Classification of Occupations (ISCO-08) issued by the ILO, explicitly considered as an integral part of the health workforce alongside those occupations classified in Sub-major group 22: Health Professionals (International Labour Organization, 2012). Thus, it stands to reason that formal career paths and opportunities for professional development of clinical engineers should be similar, if not the same, as career paths and opportunities for other health professionals. There are certainly reasons why health professionals attend specific schools, colleges, and universities, attend internships, residency (specialization), and fellowship (sub-specialization) programs, go for board certifications, and carry on with continuing professional education. It makes sense for clinical engineers to follow the pattern set by other regulated health professions and to create approval processes to ensure that clinical engineers continue to support and advance patient care by applying engineering and managerial skills to healthcare technologies (Medvedec & Yadin, 2011, Medvedec, 2014, Nagel, 2009).

One of the professional approval processes in clinical engineering is a certification. Since people may confuse terms like registration, certification, accreditation, credentialing, and

Clinical Engineering. DOI: http://dx.doi.org/10.1016/B978-0-12-803767-6.00014-3

197

licensing, while some of the terms are occasionally used interchangeably, it may be worth highlighting definitions, similarities, and differences among these terms within the professional context (Medvedec & Yadin, 2011, Medvedec, 2014).

Credentialing is an umbrella term used for different approval processes including accreditation, certification, licensure, and registration. Credential is defined as an attestation of qualification, competence, or authority of professionals issued to individuals by a third party with an authority to do so. Obtained professional credential demonstrates proficiency in the field of interest and identifies individuals who are committed to their profession, and provides assessment and recognition of their background, experience, and legitimacy to meet predetermined and standardized criteria.

Certification is generally a third-party attestation where specified requirements related to persons, products, processes, or systems have been fulfilled. In order to apply professional standards, increase the level of practice, and protect the public, a professional organization may establish a certification. Professional certification earned by an individual to perform a job or task is often called simply certification. In this context but in other words, certification is the process of issuing a certificate—a statement or declaration such as diploma, degree, title, clearance, etc.—formally attesting that a knowledge, know-how, skills, and competences acquired by an individual have been assessed and validated by a competent body against predefined standards. Professional certifications may further require certain work experience in the related field before the certification is awarded, either for a lifetime or as a time-limited recognition of an individual. Certifications are usually earned from professional societies, but also from universities and private certifiers for some specific certifications. Certifications are very common in the healthcare sector and are often offered by particular specialties. An example of such a certification process is a physician who receives certification by a professional specialty board in the practice of, for instance, radiology. The most general type of certification is profession-wide and is intended to be portable to all places where a certified professional might work. Certification is a voluntary process and it is based on the premise that there is a right to work. However, it is not a permission to work, but rather a statement of completion or qualification, with the purpose to educate and inform. Certification may be withdrawn at any time by the issuing organization, but this does not stop one from working. Licensure and certification are similar in that they both require the achievement of a certain professional level.

Accreditation is a third-party attestation related to a conformity assessment body conveying formal demonstration of its competence to carry out specific conformity assessment tasks. It is a formal process of quality assurance through which accredited status is granted, showing it has been approved by the relevant legislative or professional authorities by having met predetermined standards. Accreditation standards are usually regarded as optimal and achievable, and are designed to encourage continuous improvement efforts within accredited organizations. Accreditation is often a voluntary process in which organizations choose to participate, rather than the process required by laws and regulations. It is common in the healthcare sector, and an accreditation decision about a specific healthcare organization is usually made following a periodic on-site evaluation by a team of peer reviewers, typically conducted every few years. Although the terms accreditation and certification are used interchangeably, accreditation usually applies only to organizations, while certification may apply to individuals, as well as to organizations. When applied to individual practitioners, certification usually implies that the individual has received additional education and training, and demonstrated competence in a specialty area beyond the minimum requirements set for licensure. When applied to an organization, or part of an organization, such as the laboratory, certification usually implies that the organization has additional services, technology, or capacity beyond those found in similar organizations.

Licensing is generally a mandatory approval process by which a governmental authority grants time-limited permission (license) to individuals or organizations, after verifying that they met predetermined and standardized criteria, to perform an activity that is otherwise forbidden but considered to be hazardous, or which requires a high level of expertise. Licensing presumes that the engagement in the particular field is a privilege rather than a right, so the given privilege may be withdrawn at any time by the issuing authority. The purpose of licensing is to restrict entry and strictly control a profession or activity by ensuring that the licensee has met eligibility requirements and passed some form of assessment, usually at the state level and required by law. The license may be renewed periodically through payment of a fee or proof of continuing professional development, by inspection, etc. Licensure is common in medicine, nursing, pharmacy, psychology, social work, engineering, etc., but hardly ever in clinical engineering. The main aim of that licensing is to protect public health and ensure patients' safety. Professional associations are often an important resource and support to those looking to obtain a special level of certification or licensure.

Registration means that as a professional one has to get inserted into an official register organized by a regulatory body, usually by recording or registering the certificates. This register has standards for training, professional skills, behavior, health, etc., which registrants must meet in order to become registered and must continue to meet in order to maintain their registration or license. In most cases, the terms "licensure" and "registration" are also used interchangeably.

International Register of Clinical Engineers

The International Federation for Medical Electronics and Biological Engineering was founded in 1959. In the mid-1960s, the name of that organization was shortened to the present name the International Federation for Medical and Biological Engineering (IFMBE). As part of the organizational structure of the IFMBE are currently two special divisions: Clinical Engineering Division (CED/IFMBE), and Healthcare Technology Assessment Division (HTAD/IFMBE). Originally established as a working group in 1979, the CED/IFMBE attained official division status in 1985.

In 1981, the Agreement on Mutual Recognition of Qualifications for Clinical Engineers was signed by 22 Affiliated National Societies (ANS) of the IFMBE (Austria, Australia, Belgium, Canada, Denmark, Finland, Federal Republic of Germany, France, German Democratic Republic, Hungary, Israel, Italy, Japan, Mexico, Netherlands, Norway, Spain,

South Africa, Sweden, the United Kingdom, the United States, and Yugoslavia), mutually agreeing to recognize any holder of the IFMBE's Certificate of Registration as a Clinical Engineer, subject only to such additional criteria as might be specified by each ANS in addition to that document (Biomedea, 2013).

Mechanisms of registration were developed and elaborated as follows. The International Registration Board (IRB) is responsible for the Registration of Clinical Engineers (Nagel, 2009). The IRB consists of the Chairman of the National Examining Authority (NEA) from each of the ANS party to the Agreement, plus representatives of independent international bodies and others as appropriate. Each ANS establishes the NEA in each country, and acts as a communication channel with the IRB. The NEA comprises (grandfathered) clinical engineering professionals or those professionally capable of fully understanding the role, importance, and requirements of clinical engineering. The NEA in each country recommends individual candidates to the IRB for registration conferred by the IRB. CED/IFMBE establishes the Constitution and By-Laws of the IRB to be approved by the IFMBE Administrative Council. A nonrefundable fee for certification covers the cost of processing applications. Each NEA publishes operational guidelines, submits its Constitution and By-Laws to be approved by the IRB, takes care about funding, participates in IRB's activities through its Chairman, organizes collection and processing of applications, sets up and conducts examinations for candidates, and recommends actions to the IRB. The exact form of the examination process (written, oral, view, etc.) is left to the individual NEAs, but has to satisfy the requirements of the IRB, that is has to be meaningful so that successful candidates perform well within the specialty without preventing well-qualified individuals from attaining certification.

In order to obtain international registration as a clinical engineer, the Agreement defined that a candidate must have successfully completed a basic education in engineering or applied sciences to a BSc level and have had not less than 3 years of pertinent clinical engineering experience, or, in addition to achieving that BSc education, to have MSc or PhD education and/or training in biomedical engineering and have had not less than 2 years of pertinent clinical engineering experience. NEAs may, at their discretion, but with the approval of the IRB, impose additional requirements as may be dictated by local national practices. Two years of relevant experience counts as 1 year of training, where experience is offered instead of training.

Formally, this Agreement seems to be still in place, but since registration and/or certification have never been made mandatory by national legislations in most of the countries, the agreement has been neglected and the project of international registration and/or certification has actually never been accomplished to a full extent using defined mechanisms, as elaborated above. One of the possible initial project attempts appeared to be the first issue of the International directory of clinical engineers released in 1994, containing names of more than 1200 individuals from 62 countries, with the intention to improve recognition, communication, and networking within global clinical engineering community.

Though neither the laws nor employers usually ask for certificates in clinical engineering, there is a significant interest in and need for clinical engineering certification, as shown by the latest global CED/IFMBE survey among clinical engineering professionals and associations. Biomedicine and healthcare challenges in the twenty-first century urge regulation of all health professions worldwide, so one international umbrella program for certification in clinical engineering is among the priorities of the global clinical engineering community. Currently ongoing CED/IFMBE project on international clinical engineering certification is expected to help in finally achieving this prioritized goal.

Certification in the United States

Certified Biomedical Equipment Technicians

Certification in the United States in the clinical engineering field began with biomedical equipment technicians (BMETs) (Croswell, 1995). As biomedical equipment maintenance was developing in US hospitals in the late 1960s, there were no training programs for BMETs. A few 2-year technical schools initiated training programs based on their electronics curriculum, but there was no standardized curriculum. Most BMETs had been trained in the military schools.

The Association for the Advancement of Medical Instrumentation (AAMI) established a task force to look at the BMET field and the maintenance of medical equipment in hospitals. The task force decided that certification was needed to allow BMETs to demonstrate that they had a minimum level of expertise. A Board of Examiners was established by AAMI and the first exam was given in 1970. Individuals who passed the written exam became CBETs.

In 1973, the Department of Veterans Affairs (VA) developed its BMET certification program from its Engineering Training Center (Wear & Van Noy, 1979). The requirements to take the exam were the same as the AAMI and the exam was similar since the Director of the VA Training Center was on the AAMI Board of Examiners. The VA exam was developed by the VA Training Center faculty. The VA technicians were called biomedical engineering technicians, which is still BMET.

AAMI also developed specialist exams for BMETs who work on laboratory and radiological equipment. These BMETs do not have to take the general exam since they only worked on special equipment, but they need certification to demonstrate a minimum level of expertise in their specialty. Individuals who pass these exams become Certified Radiological Equipment Specialists (CRESs) and Clinical Laboratory Equipment Specialists (CLESs).

In 1984–1985, the VA merged its certification program with the AAMI program. All VA certifications were accepted by the AAMI program.

Electronics Technicians Association International (ETA) also certifies BMETs as both general medical equipment and radiological equipment technicians (ETA Senior and Master Certifications, 2013). They have to be certified as Certified Electronics Technicians (CET) before they can take the journeyman certification exams. With 6 or more years of training and work experience in the field, the CET can take the journeyman exam (ETA Journeyman certifications, 2013). They must score 85% on the journeyman exam to be certified. If they pass the journeyman certification exam for medical equipment, they become CET-BMD. By passing the journeyman certification exam for (radiological) imaging equipment, they become CET-BIET. Their programs are aligned with the ISO/IEC 17024 standards *"Conformity assessment—General requirements for bodies operating certification of persons."*

Individuals must meet the following qualifications to take the AAMI-BMET certification exam: have an associate degree in BMET and 2 years' experience full time as a BMET, or have an associate degree in electronics technology and 3 ½ years' experience full time as a BMET, or have 4 years' experience full time as a BMET.

The exam can be taken if a person has an associate degree in BMET or 2 years' experience full time as a BMET. If they pass the exam, they have 5 years to complete the additional 2 years of full-time experience as a BMET to be certified. To take the CRES or CLES exam, a person must have worked at least 40% of the time in the past 2 years or 25% of the time in the past 5 years in the designated specialty area.

Each of the AAMI exams has 150 multiple-choice questions and is administered by a professional testing organization. The Board of Examiners creates questions for the exam bank and reviews new exams before they are used. The professional testing organization has responsibility for the exam security.

In the AAMI certification program over 6000 are CBETs, about 600 CRESs, and around 100 CLESs. Every 3 years, individuals must renew their certification by demonstrating a certain amount of continuing education to be maintained as a CBET, CRES, or CLES. Most of the CBETs are in the United States, but BMETs in several other countries have taken the exam and become CBETs.

Certified in Clinical Engineering (CCE)

With the increase of medical instrumentation in the hospital, more of the personnel performing engineering-type work in the clinical setting in hospitals came into the field. Also as BMETs were hired to maintain the medical instrumentation, engineers and physical scientists were hired to manage clinical/biomedical engineering departments in hospitals.

AAMI developed the first certification program for clinical engineering in the United States. Since several of the major people working in the field of clinical engineering were not engineers, it was decided that a program should be developed to certify people in the clinical engineering field but not as clinical engineers. An initial Board of Examiners was established with prominent people in the clinical engineering field. The Board decided that for 1 year, individuals working in clinical engineering could be certified based on credentials. They had to have at least a BSc degree in engineering or a physical science and at least 3 years' experience working in the field of clinical engineering. These credentials were evaluated by the Board of Examiners. The AAMI certification program for clinical engineering was established in 1975 and within a year about 200 individuals were "certified in clinical engineering" (CCE).

After this first group of CCEs, the Board of Examiners developed a written exam and an oral exam to test future individuals for certification.

At this same time, another group of prominent individuals in the field decided that people should take an exam to become certified in clinical engineering. As a result five people self-certified themselves and developed an exam for certification in clinical engineering. This group became the American Board of Clinical Engineering (ABCE) and also started their certification program in 1975. Most of the initial individuals in this program were academic clinical engineers.

AAMI and the ABCE certified individuals in clinical engineering until 1984. The two groups merged in 1984 with all the ABCE CCEs being accepted into the AAMI certification program. As part of the merger the International Certification Commission for Clinical Engineering and Biomedical Technology (ICC) was established. Fifty individuals had been certified by the ABCE.

Starting in 1979, AAMI required all CCEs to renew their certification every 3 years by demonstrating continuing education. AAMI discontinued accepting applications for certification in 1999 because there were not enough applicants to support the program financially. However, they did continue to accept renewals. At that time, there were 474 CCEs including 50 certified by the Canadian Board of Examiners for Clinical Engineering and several individuals in other countries certified by the US Board of Examiners (Nicoud & Kermit, 2004).

In 2002, the Healthcare Technology Certification Commission (HTCC) was created under the Healthcare Technology Foundation (HTF) to reestablish a CCE program. A US Board of Examiners was created to develop the written and oral exams. The written exam was based on the American College of Clinical Engineering (ACCE) Body of Knowledge (BOK) determined by an ACCE survey of practicing clinical engineers. This survey asked the clinical engineers about the work that they were doing and the knowledge requirements. The new certification program accepted anyone from a previous certification program who demonstrated that they were current in the field by continuing education for a 1-year period. One hundred and twelve (112) individuals were accepted into the new program from the ICC program. In 2004 the first exams were given.

The HTCC has created a Canadian Board of Examiners to develop exams and recommend certification of Canadian clinical engineers. In 2014, there were about 200 individuals certified by the HTCC from the US and Canadian Board of Examiners.

In 2013, the HTF determined that it could no longer sponsor HTCC because new US tax policies indicated that a nonprofit foundation such as HTF could not have an income-producing unit like HTCC. In 2014, the ACCE became the sponsoring organization for HTCC.

Individuals must meet the following qualifications to take the CCE exam: 3 years of clinical engineering experience plus Profession Engineer License, or MSc Eng or BSc Eng plus 4 years of total engineering experience, or bachelor of engineering technology (BSET) plus 8 years of total engineering experience.

They also must provide three professional references. The written exam has 150 multiple-choice questions administered by a professional testing company. The questions are developed by the Board of Examiners and are based on the Body of Knowledge developed by the ACCE. The oral exam is about 2 h and given by two members of the Board of Examiners and is based on practical knowledge needed to function on the job.

Other Programs Under the ICC and HTCC

In the 1990s, nine Brazilian clinical engineers were certified by the ICC and a Brazilian Board of Examiners for Clinical Engineering Certification was created. However a certification program was never developed (De Magalhaes Brito et al., 2004).

In 1991, the first Mexican clinical engineer was certified by the ICC and in 1993 the CCE exam was given in Mexico and three more Mexican clinical engineers were certified. In 1994, the Mexican Board of Examiners was approved by the ICC and the ICC accepted the affiliation of the Mexican Certification Commission (MCC). The MCC administers the CCE exam in Spanish (Velasquez, 2004).

Hong Kong has had their BMETs take the ICC exam used in the United States, but has not had a Hong Kong Board of Examiners. For CCE they have used the US exam process under the HTCC (Poon, 2010).

Certification in Canada

Canada uses the ICC for certifying BMETs and they add the requirement that an individual must have a BSc in biomedical technology to take the exam. The exam is developed by the Canadian Board of Examiners under the ICC.

Under the laws of the Canadian provinces and territories, the use of the title "engineer" in a job description requires that the incumbent be licensed as a professional engineer in that jurisdiction. Canada has always taken the position that to be eligible to seek certification in clinical engineering, an applicant must first obtain licensure as a professional engineer. Once a person is licensed as a professional engineer and is working in the field of clinical engineering, then he or she can apply to the Canadian Board of Examiners for Clinical Engineering Certification.

By 1980, it was recognized that engineers working in the clinical engineering role required a distinct but unrecognized body of knowledge to perform their tasks competently. Since there was no licensing process in place specifically for clinical engineering, leaders in Canada decided to establish a certification process that would be administered by competent members of the profession. In order to begin such an effort, discussions were held with colleagues in the United States who had undertaken a similar approach under the leadership of the AAMI. Canadians with established track records working in the profession were grandfathered as certified, and established the first Canadian Board of Examiners for Clinical Engineering Certification. They developed a written exam and an oral exam.

This process of certification continued for a number of years. However, the initial rush of applicants dwindled and it remained a voluntary activity with limited visibility among the healthcare community. By the late 1990s, the work of the Board had effectively ceased with very few applicants coming forward.

Around 2008, there was a growing interest in certification in Canada as younger engineers entered the profession and the need for skilled staff continued to grow. Members of the former Canadian Board were asked by the Canadian Medical and Biological Engineering Society (CMBES) to restart a Canadian certification process and bring it up to date. It was apparent that with the small number of certification applicants, it would be difficult to launch and sustain a self-supporting certification process. Since there are many similarities in the practice of clinical engineering between Canada and the United States, they decided to approach the US Board about the possibility of sharing aspects of the enhanced US exam process.

Adding further credibility to the process, The US Board of Examiners is accountable to the Health Technology Certification Commission, which oversees the work of the Board and ultimately decides on recommendations from the Board to certify individuals.

Discussions between the Canadian and US Boards went well with good support and encouragement from US colleagues. The main issue of divergence of practice between Canadian and US clinical engineers relates to the country-specific codes, standards, and regulations, an important but relatively small part of the written exam. In discussion, it was agreed that members of the Canadian Board would review the US written exam, to identify those questions requiring specific knowledge of US codes, standards, and regulations. Out of a full exam of 150 multiple-choice questions, the total number of exempted questions is typically no more than 30. These questions are not counted for Canadian examinees and the same percentage pass mark is used. To compensate for the lack of written exam questions on Canadian codes, standards, and regulations, it was decided to put an additional (fourth) question into the Canadian oral exam process, specifically on these topics. The Canadian Board agreed to develop such a question using exactly the same process as the US Board. In this way, Canadian candidates are examined through a slightly different but parallel process to their US counterparts.

It was agreed that Canadian applicants would register and be administered by the Secretariat to the US Board, to avoid setting up a parallel office in Canada. Sites are available in Canada to sit for the written exam, which is made available in both countries on a single date and time each year, early in November. All policies and procedures are harmonized and the Canadian Board assists the US Board in the generation of new written and oral exam questions. Members of the two Boards discuss their work on a regular basis, and the Chairs of each Board sit on the Healthcare Technology Certification Commission.

The harmonized process was established in 2010 and remains in place. There has been good communication between each Board, and a generally high level of support for this harmonized process (Easty, 2015).

Commission for the Advancement of Healthcare Technology Management in Asia (CAHTMA) and Certification

CAHTMA was initiated in 2005 with the endorsement of the Asian Hospital Federation (AHF) (Wear, 2012). AHF is an international nongovernmental organization, supported by members from 14 countries in the Asia Pacific Region. CAHTMA is a member of the IFMBE. It was established to provide a platform for healthcare professionals to discuss and exchange ideas on healthcare technologies and practices. Central to these objectives are the promotion of best technology management practices, the certification of clinical engineering practitioners and healthcare professionals, and the dissemination of appropriate management tools through seminars and workshops.

CAHTMA has certified a few clinical practitioners, but there has not been a major need for certification in Malaysia since it has not been required. Technicians are certified as level-one clinical practitioners with a written exam and experience which is similar to the ICC BMET. Engineers are certified as level-two clinical practitioners with a written exam and an oral exam and experience which is similar to the HTCC CCE. To encourage more engineers to become certified, CAHTMA is going to use the process of certifying individuals based on credentials similar to what has been done with the initial program in the United States and Taiwan.

CAHTMA is also certifying faculty for biomedical engineering technology programs which are developing with the increased need for technologists to maintain the medical equipment. In 2012, lecturers at one school were tested as assessors and certified by CAHTMA with Certification for Clinical Engineering Assessors. Lecturers who completed 5 weeks of training and passed the exams were certified by CAHTMA with Certification for Clinical Engineering Trainers.

In 2015, CAHTMA is planning to start certifying clinical engineering practitioners in India.

Certification in Taiwan

The Taiwan Society for Biomedical Engineering (TSBME) performs certification in clinical engineering in Taiwan (Chang and Lin, 2010). In 2000, TSBME established the Certification Executive Committee for CE certification. During 2001, they certified clinical engineers by application. In 2003, a recertification program for CCE was initiated. The first testing for certification of clinical engineers and technologists of medical equipment was in 2007.

The TSBME provides certification for clinical engineers, medical equipment technicians, and biomedical engineers. In 2010 they had certified 93 clinical engineers, 132 medical equipment technicians, and 224 biomedical engineers. The clinical engineers and medical equipment technicians are for working in the hospitals and the biomedical engineers are for working in the medical device industry. This is the only certification that has a separate certification program for hospital and industry engineers.

To become certified an individual must be a member of TSBME. The requirements to take the certification exam are as follows:

Clinical Engineer: MSc degree in biomedical or related field plus at least 1 year of CE experience plus working in a hospital for more than 10 years.
Medical Equipment Technician: BSc degree in biomedical or related field plus at least 1 year of CE experience plus working in a hospital for more than 4 years.
Biomedical Engineer: BSs degree in Engineering plus at least 2 years of BME experience plus working in BME field for more than 4 years.

The content of the assessment exams by the TSBME for each of their certifications is as follows:

Clinical Engineer (core exam plus oral exam)

Anatomy (24%)

Medical Instrumentation (16%)

Clinical Engineering (16%)

Medical Imaging System (16%)

Major Area (Biomechanical or Biomaterial or Medical Electronics or Medical Information (28%))

Medical Equipment Technician (core exam)

Anatomy (20%)

Electronics & Electrical Safety (40%)

Medical Instrumentation (40%)

Biomedical Engineer (core exam)

Anatomy (20%)

Medical Devices, Safety Regulation & GMP (10%)

Major/Minor (Biomechanics plus Biomaterial or Medical Electronics plus Medical Instrumentation) Major 45% and Minor 25% (70%)

Certification in Japan

Clinical engineering in Japan is different since it is the country where the government certifies clinical engineering technologists (CETs) (Kanai, 2004 & Umimoto, 2013). The CETs must graduate from a clinical engineering training school which can be a university, junior college, or training school and pass a national exam to be certified. The CETs are also called clinical engineers. The CETs are paramedical staff and specialize in the medical equipment essentials in medical care. About 35% work in hemodialysis and about 20% in maintenance. Others work in respiratory, operating room, intensive care unit, heart related, hyperbaric, and other areas.

In 1987 the clinical engineering system was established by the Clinical Engineers Act. This act created the CET as a professional medical position responsible for the operation and maintenance of life-support systems under the direction of doctors. This act established a national qualification including passage of the 180 questions exam in medicine, engineering, and medical technology. In 2010 there were about 28,000 certified CETs and about 18,000 current working in the field. The certification of the CETs is most equivalent to the CBET in the ICC system in the United States.

In addition to the CET certification by the government, the Japan Society for Medical and Biological Engineering (JSMBE) has a Biomedical Engineering Certificate program (Kanai, 2004 & Umimoto, 2013). The JSMBE has two classes of certification for biomedical engineers. The 1st class certification is for experienced clinical engineers and in 2008 the pass rate was 22.2% for 433 applicants. The 1st class exam covers basic aspects on medical engineering and medical device-related subjects. The 2nd class exam is for students or recent graduates of clinical engineering and many take it as preparation for the national CET exam. In 2008 the pass rate on the 2nd class exam was 29.3% for 1398 applicants.

Certification in China

In 2005, the international clinical engineer certification was introduced in China. The Medical Engineering Division of the Chinese Medical Association hosted the first international clinical engineering certification training courses and certification examination (Zhou & Ying, 2013). From 2005 to 2014, there have been seven sessions of lectures by international senior specialists and certification exams. The written exam is based on the ACCE Body of Knowledge with some adjustment for the practice of clinical engineering in China. The written exam is in English and is prepared by international senior specialists. Individuals who pass this 100-questions multiple-choice exam have to pass an oral exam in English to become certified. The oral exam is given by the international senior specialists. In the seven training sessions, there have been 760 clinical engineering personnel from hospitals and universities. There have been 252 individuals who have passed the two exams and been certified as international clinical engineers. The candidates for the International Clinical Engineering Certification are mostly senior clinical engineers with more than 10 years' experience.

In 2012, the Medical Engineering Division of the Chinese Medical Association carried out Chinese Registered Clinical Engineer Certification (RCEC) training and examination. The candidates were junior engineers in large hospitals or new graduates with majors in medical engineering. This exam is the basic admission exam to the occupational qualification of clinical engineering.

The RCEC exam consists of a theoretical exam and practical test. There is a Chinese exam question bank from which the theoretical questions are randomly selected. Candidates then take a practical test including repair, measurement, and maintenance of medical devices. A committee of Chinese clinical engineering experts evaluates the ability of the candidates and determines if they are qualified to receive the RCEC. In 2013, there were 176 people who took the exam and 56 passed to become certified as RCEC. In 2014 an additional 46 were certified for a total of 102 RCECs.

They are establishing a continuing education for both certifications to maintain and improve the quality of the clinical engineers. The Medical Engineering Division plans to recommend to the government to officially authorize clinical engineer training and certification.

Certification in Sweden

The Swedish Society for Medical Engineering and Physics (MTF) started the Certification of clinical engineers in 1994 (Wandel et al., 2015). The certification is performed at two levels, by an examination corresponding to approximately the level of a Bachelor's degree in engineering, and by an examination corresponding to approximately the level of a Master's degree in engineering.

Applications can be sent to the Society twice a year. They are judged by a Certification Committee that has a mandate to review the applications from the Board of the Society. The Certification Committee consists of a Chairman who preferably is a lawyer from a governmental healthcare organization or a healthcare provider. There are two university professors in biomedical engineering, and two experienced certified clinical engineers. There are also deputy members who are certified clinical engineers.

The requirements, besides the university certificate on the candidate's engineering examination, are courses in biomedical or clinical engineering, medicine, and related subjects corresponding to at least 30 credit points. The European Credit Transfer and Accumulation System (ECTS) is used as a reference. The courses can be university courses or courses given by other organizations or by companies. Credit points are assigned to each course by the Committee.

Different types of courses can be approved. One type is courses in medicine or biomedical engineering, and the credit points from such courses should be at least 15 points. The certified engineer should also have at least eight credit points in specific clinical engineering subjects such as technical safety in health care, clinical engineering management, quality assurance, or risk analysis.

One could also have continuing education in engineering subjects with direct relevance for the work as a clinical engineer. This can correspond to computer science courses, electronics, fluid mechanics, measurement science etc. Fifteen credit points in biomedical or clinical engineering can come from courses included in the basic examination. Furthermore, at least 15 credit points should be part of post-examination continuing education. To become a certified clinical engineer, the person should have at least 3 years of work experience as a clinical engineer in a hospital supervised by an experienced and preferably certified clinical engineer.

The reason for choosing the Bachelor of engineering as the lower level of certification was that in 1989, a Swedish law came into force, which stated that to work with clinical engineering, a person should preferably have at least this level of education. At the time, there were many engineers working at the clinical engineering departments in the hospitals who did not have the Bachelor's degree but an older degree from a polytechnic institute. These engineers were accepted for certification if the degree was from 1989 or earlier. However, they had to have at least 6 years instead of 3 years' work experience.

Since 1994, there have been a total of 695 applications, 124 at the Master's level and 571 at the Bachelor's level. A total of 391 persons have been certified of whom 87 are at the Master's level and 304 at the Bachelor's level.

A program to certify specialists in clinical engineering was developed in 2014. To become a certified specialist in clinical engineering, the engineer should have at least 2 years in specialist training supervised by an experienced and certified specialist in clinical engineering. The specialist training program consists of courses corresponding to a Continuing Professional Education. It is the Certification Committee that classifies the courses for the specialists. The certified specialist should have at least 30 credit points during the specialist training years. The specialist certification is also performed at two levels (a Bachelor's degree and Master's degree in engineering, respectively). To keep the role of a specialist they should continue to develop their professionalism as a clinical engineer by ongoing training and education.

Specialist programs for a number of different types of medical functions are under development. Examples of Specialist programs are Medical Imaging, Dialysis, Intensive Care, Computers in Health Care, Responsibility, and Management.

Certification in Poland

In 2002 the program of specialization in medical engineering for engineers as professionals in clinical environment was introduced by the national legislation, under the auspices of the Ministry of Health, the Medical Centre of Postgraduate Education, and the National Consultant in the field of medical engineering, in a way similar to education and training programs for medical professionals (Zalewska et al., 2014). The candidates for this specialization must have MSc degree in biomedical engineering, automatics and robotics, electronics and telecommunications, mechanical engineering, or computer science, and have at least 3 years of work experience in clinical environment. The workload of this postgraduate medical engineering specialization program is 1700 hours during 2–3 years, with about half filled with lectures and laboratory exercises, and the rest practicing in hospitals and clinics having appropriate facility, equipment, and staff for such activity, and being accredited by the State Commission for Accreditation. Training program modules are:

1. Basic Medical Knowledge
2. Biomechanics and Rehabilitation Engineering
3. Fundamentals of Medical Electronics
4. Radiological Devices and Radiation Protection
5. Automatics, Robotics, and Healthcare Telematics
6. Signal Processing, Modeling, and Medical Informatics
7. Electrography, Intensive Care Instrumentation, and Laboratory Equipment
8. Computer Tomography (XCT, MRI, SPECT, PET) and Ultrasonography
9. Biomaterials and Artificial Organs
10. Clinical Engineering Regulatory and Organization.

At the end of the program, in order to obtain the title and the certification as a specialist in medical engineering, the candidate has to pass the practical and theoretical part of the state exam in front of the State Examination Commission. Professional competences gained during postgraduate education, entitles successful individuals for work in clinical environment as a medical engineer. Moreover, a participation of certified specialists in medical engineering in some advanced medical procedures is also required by law, as well as the positions of National and Regional Consultants for medical engineering issues.

Certification in Czech Republic

Since 2004 the Act on Nonmedical Health Service Occupations and the related regulations recognize health service professionals with technical competences (biomedical technicians with BSc degree in biomedical technology, and biomedical engineers with MSc degree in biomedical engineering), and health service specialists with specialized technical competences (clinical technicians with BSc in clinical technology, clinical engineers with MSc degree in clinical engineering), as a new part of similar legislation that have already existed for decades for medical professionals (Lhotská & Cmíral, 2009). New legislation regulates details about holding the attestation examination, examination for issuing the certificate, final examinations of accredited qualification courses, approbation examinations, and examination rules for these examinations; concerns health capabilities for performing the occupations of health service professional and other specialists; defines credit system for issuing certificates to performing health service occupations without expert supervision; defines credit points educational activities; specifies activities of health service professionals and other employees; defines fields of specialized education and specification of the expertise of health service professionals; specifies minimum requirements for educational programs leading to qualification to act as health service professional, etc. If professionals with technical competences come into contact with patients or if they can directly impact the patient's health through their professional activities, they are required to have qualification of either health service professional or health service specialist. There is an established system of undergraduate (BSc, MSc), postgraduate (PhD), specialized and lifelong education and courses, accredited by ministries of education and health, to gain such qualifications and appropriate certifications. The graduates of accredited study programs and fields get the certificate of qualification to perform health service occupations. Graduates of other bachelor or master study programs in electrical engineering can obtain the qualification for health service professionals with technical competence if they complete the course in Biomedical Engineering (MSc Eng) or Biomedical Technology (BSc) that are accredited by the Ministry of Health Care. Graduates of other study programs must complete specialized postgraduate courses in biomedical engineering. Apart from theoretical courses in medical fundamentals (anatomy, physiology, pathology, pathophysiology) and in technical area (mathematics, physics, medical devices, signal theory, medical imaging, databases, etc.), an inseparable part of the education is practical training in healthcare facilities. The specialized technical competences for clinical technicians and clinical engineers can be obtained by completing specialized education and training and passing official examinations in front of the board appointed by the Ministry of Health Care. The board members are appointed by the Minister of Health Care based on the proposal of professional societies, associations, and accredited institutions. This specialized education and training can be provided only by those institutions that have the accreditation from the Ministry of Health Care. Clinical Engineering as specialized education and training for Biomedical Engineers and Clinical Technology for the Biomedical Technicians are types of education organized by the Institute for Postgraduate Education in Health Care directly controlled by the Ministry of Health Care. This education includes several fields, such as signal acquisition and signal processing, diagnostic devices, laboratory devices, therapeutic devices, diagnostic imaging devices, and perfusiology. Biomedical Technicians and Biomedical Engineers can also complete specialized education in the Health Service Organization and Control. After the official examination they are ranked as Professionals in Health Service Organization and Control. For Health Service professionals (Biomedical Technicians and Biomedical Engineers) and specialists (Clinical

Technicians and Clinical Engineers) the Ministry of Health Care issues the official certificate. Then the professionals can apply for registration in the Registry of Healthcare Professionals which is a part of the National Health Care Information System. They receive a certificate that is valid for 6 years and then they have to renew the registration. The basic conditions are: healthcare practice during the last 6 years (minimum 1 year 0.5-part-time job or 2 years 0.2-part-time job); acquisition of minimum 40 credit point through participation in different activities during the last 6 years. Each activity must be recorded in the personal certificate of specialization. This information serves as a basis for application of registration or registration renewal. As a part of lifelong education, a credit system was introduced that specifies activities for which credit points can be awarded.

Certification in Italy

In Italy, a process of defining common rules for recognizing the activities of a biomedical and clinical engineer and for the certification of the skills of engineers is currently in progress (Iadanza, 2015).

The Territorial Associations of Professional Engineers have the right to set up voluntary certification of skills for their members.

The Italian laws in recent years explicitly have mentioned clinical engineers and clinical engineering services, making even more urgent the need for a certification procedure.

The laws state that the Territorial Associations of Professional Engineers are responsible for defining a set of rules for certification. The document, being drafted by the local biomedical engineering committees, will identify a metric of evaluation that is based on the verification of the contents of the documents submitted for the recognition of skills, on the interview with the candidate and on the evidence for continuity of professional activity.

The candidate must provide the following documentation:

- Title for Degree in Biomedical Engineering or other Engineering Degrees with biomedical orientation. In this case, the registration will be on the activities of Section A (the section reserved to professionals having a 5 years "Laurea")
- Title for Bachelor's degree in Biomedical Engineering. In this case, the approval will be on the activities of the Section B (the section reserved to professionals having a 3 years "Laurea")
- Successful passage of the State Exam
- Active Enrollment in the Territorial Association of Professional Engineers
- Curriculum Vitae with documented activity inherent in the specific area of specialization for Biomedical Engineer and, if any, postgraduate Master's degree in any inherent discipline (e.g., Clinical Engineering, Risk Management, Technology Assessment).

The qualifying activities for a Clinical Engineer are identified as follows:

- Planning, purchases, and replacements evaluation, Health Technology Assessment (HTA)
- Perform installation, testing, and control of safety and quality
- Management of maintenance activities
- Management activities relating to safety for medical devices and biomedical technologies

- Performing maintenance activities
- Training and consultancy
- Standards and standardization
- Risk Management for medical devices and biomedical technologies in a hospital setting
- Integration of medical devices, including medical software, within medical information technology networks
- Participation in ethics committees.

Certification in Ireland

In 2003, in anticipation of forthcoming legislation for Statutory Registration of Health and Social Care Professions, the Biomedical Engineering Association of Ireland (BEAI) and Biomedical Engineering Division of Engineers Ireland established a Clinical Engineering Voluntary Registration Board (CEVRB) and an associated Clinical Engineering Registration Scheme, as a voluntary professional registration plan (Smith et al., 2015). The CEVRB was composed of Engineers from academia, practitioners from the public and private sector, and representatives of publically and privately funded hospitals. The plan, which was developed, considered Education, Clinical Engineering Experience, Ethics, Professional Standing, and Continuing Professional Development (CPD).

Developed plan is based on achieving Engineers Ireland's Professional Titles (Engineering Technician, Associate Engineer, and Chartered Engineer). Engineers Ireland has statutory responsibility for the title of Chartered Engineer in Ireland. The three protected titles of Engineers Ireland require the achievement of a specified academic standard, a specified minimum number of years of experience, an interview based on a set of published competencies, and an engagement with a code of ethics.

Engineers Ireland also has a well-developed plan to support "grandfathering" with a process which recognizes experience in lieu of academic qualifications by assessing the candidate's ability with respect to specific competencies.

In addition, the plan included an application form where two recognized practitioners would sign-off on the candidate's experience in the clinical engineering field. A voluntary Continuing Professional Development scheme was also developed.

Since Clinical Engineering is a small profession in Ireland, Statutory Registration will not be implemented for some years. It is thought the Clinical Engineering Voluntary Registration plan will meet the short-term requirements for a registration plan.

Other Certifications

South Africa has a voluntary registration program of Clinical Engineers, Clinical Engineering Technologists, and Clinical Engineering Technicians, which is based on experience and academic requirements (Wear, 2008 & Wear, 2012). All three of these groups are considered professionals and would have academic training. They also have medical equipment repair personnel.

Germany has developed a Certified Clinical Engineering program, but it does not require an exam. It is based on experience and academic background. They are also planning to develop a certified Biomedical Engineering Technician program. In most European countries, there are more engineers than technicians and so the certification of the engineer is more important.

The United Kingdom initially developed a certification for Clinical Engineers in the 1990s but this program has been dropped due to lack of interest in it. They are now having a voluntary registration for Clinical Engineering Technologists, which includes Clinical Engineers, Medical Physics individuals, and other scientific people who will be working in the clinical setting.

References

Biomedea, 2013. Available from: <http://www.biomedea.org/>.

Chang, W.H., Lin, K.P., 2010. Development and progress of biomedical engineering certification programs in Taiwan. In: Workshop on clinical engineering education and training. Workshop conducted at the 3rd International Conference on the Development of Biomedical Engineering, Ho Chi Minh City, Vietnam.

Croswell, D.W., 1995. The evolution of biomedical equipment technology. J. Clin. Eng. 20 (3), pp. 230−234.

De Magalhaes Brito, L.F., 2004. Clinical engineering in Brazil. In: Dyro, J.F. (Ed.), Clinical Engineering Handbook. Elsevier, Burlington, MA, pp. 69−72.

Easty, A.C., 2015. The evolution of clinical engineering certification in Canada. IFMBE Proceedings, vol. 45, pp. 950−953.

ETA Journeyman certifications, 2013. Available from: <http://eta-i.org/journeyman_certifications.html>.

ETA Senior and Master Certifications, 2013. Available from: <http://eta-i.org/senior_master_certifications.html>.

Iadanza, E., 2015. Clinical engineering certification and education in Italy. IFMBE Proceedings 45, pp. 954−957.

International Labour Organization, 2012. International Standard Classification of Occupations—ISCO-08 Volume I: Structure, group definitions and correspondence tables, April 2015. Available from: <http://www.ilo.org/wcmsp5/groups/public/---dgreports/---dcomm/---publ/documents/publication/wcms_172572.pdf>.

Kanai, H., 2004. Clinical engineering in Japan. In: Dyro, J.F. (Ed.), Clinical Engineering Handbook. Elsevier, Burlington, MA, pp. 91−92.

Lhotská, L., Cmíral, J., 2009. Accreditation and certification in biomedical engineering in the Czech Republic. IFMBE Proceedings 25/XII, pp. 372−375.

Medvedec, M. 2014. Global program for certification of local clinical engineers: back to the future. IFMBE Proceedings, vol. 41, pp. 1092−1095.

Medvedec, M., Yadin, D., 2011. Clinical engineering certification. IFMBE News No. 88, pp. 14−16.

Nagel, J.H., 2009. The regulation of the clinical engineering profession as an important contribution to quality assurance in health care. IFMBE Proceedings 25/XII, pp. 376−378.

Nicoud, T., Kermit, E., 2004. Clinical engineering certification in the United States. In: Dyro, J.F. (Ed.), Clinical Engineering Handbook. Elsevier, Burlington, MA, pp. 617−618.

Poon, A., 2010. Development of biomedical engineering in Hong Kong—from cradle to certification. In: Workshop on clinical engineering education and training. Workshop conducted at the 3rd International Conference on the Development of Biomedical Engineering, Ho Chi Minh City, Vietnam.

Sakuma, I., Ishihara, K., 2010. JSMBE (Japan Society for Medical and Biological Engineering) activities in clinical engineering education. In Workshop on clinical engineering education and training. Workshop conducted at the 3rd International Conference on the Development of Biomedical Engineering, Ho Chi Minh City, Vietnam.

Smith, M., Mahady, J., Grainger, P., 2015. Certification of clinical biomedical engineers—the Irish experience. IFMBE Proceedings, vol. 45, pp. 958—960.

Umimoto, K., 2013. Development of clinical engineering in Japan. In: Short Paper No. 0309 at the 35th Annual conference IEEE Engineering in Medicine and Biology Society, Osaka, Japan.

Velasquez, A., 2004. Clinical engineering in Mexico. In: Dyro, J.F. (Ed.), Clinical Engineering Handbook. Elsevier, Burlington, MA, pp. 80—83.

Wandel, B., Peterson, N., Ask, P., 2015. Certification of clinical engineers in Sweden. IFMBE Proceedings, vol. 45, pp. 961—963.

Wear, J.O., 2008. Certification of biomedical engineering technicians and clinical engineers: Important or not. 7th Asian-Pacific Conference on Medical and Biological Engineering IFMBE Proceedings, vol. 19, pp. 558—561.

Wear, J.O., 2012. Global perspectives on competency certification of medical electronic graduates in the fast changing healthcare engineering field. International Keynote, Symposium on Innovation and Commercialization for Medical Electronic Technology, Bandor Enstek, Nilai, Malaysia.

Wear, J.O., Van Noy, F.E., 1979. Veterans administration BMET certification exam. J. Clin. Eng. 4 (4), pp. 311—313.

Zalewska, E., Paøko, T., Pawlicki, G., 2014. Medical engineering in Poland, IFMBE Proceedings, vol. 45, pp. 967—969.

Zhou, D., Ying, J., 2013. Development of clinical engineer certification in China. *IFMBE News* No. 93, pp. 25—27.

Index

Note: Page numbers followed by "*b*," "*f*," and "*t*" refer to boxes, figures, and tables, respectively.

Printed in the United States
By Bookmasters